**Atomic
Spectrometric
Analysis of
Heavy-Metal
Pollutants
in Water**

Winter sampling through seven feet of ice on Paxon Lake in the Alaska Range.

Atomic Spectrometric Analysis
of Heavy-Metal Pollutants in Water

DAVID C. BURRELL, PhD

Institute of Marine Science
University of Alaska
Fairbanks, Alaska

ann arbor science PUBLISHERS INC.

POST OFFICE BOX 1425 • ANN ARBOR, MICHIGAN 48106

Second Printing, 1975

Copyright © 1974 by Ann Arbor Science Publishers, Inc.
P.O. Box 1425, Ann Arbor, Michigan 48106

Library of Congress Catalog Card Number 73-92513
ISBN 0-250-40052-9

"Knowledge is of two kinds. We know a subject ourselves, or we know when we can find information upon it."

Dr. Johnson

For Helen

PREFACE

This book is addressed primarily to the working analytical chemist charged with the responsibility of evaluating the distribution of heavy metals present in "natural" bodies of water. This includes, of course, utilized and "stressed" waterways--any water, in fact, in which the metals are present in trace quantities and for which the basic water properties have not been drastically modified, such as in the case of raw industrial or municipal effluent. The contents should also be of interest to ecologists, workers and students in a host of related disciplines.

At the time of this writing, there is considerable interest and apprehension about the role and fate of heavy metals in general in the waters of the more advanced industrial nations. And, while it is true that this concern currently centers upon a few elements only (of which mercury, lead, cadmium and arsenic are the most important), technological advances continually bring other candidates to the fore. Much of the present uncertainty as to actual and potential deleterious effects of these metals may be directly attributed to the paucity of available information on the concentrations and pathways within the natural and stressed environments. It is only within the previous decade or so that reliable methods for determining the aqueous trace metals have become widely available, and several examples could be cited in which apparent "problems" have first come to light following some advance in analytical technique. It is more than likely that additional elements are waiting to capture the pollution headlines. One of the principal objectives of this book, therefore, is to consider the analytical aspects of potential metal pollutants as but one facet--albeit an exceedingly important one--of the total environmental picture. All too often a gulf exists between the analysts on one hand and the biologists, physical chemists *et alia* whose parish is an understanding of all the inorganic and organic processes which constitute the various ecosystems. This dichotomy may lead to the establishment of arbitrary and inflexible standards for the metal contents of potable waters and food materials--with all that this implies in terms of economic stresses-- or even to misuses and inadmissible manipulations of the data. One such example of the problems inherent in divorcing the disciplines seems to me to be a tendency to relate the analytical results to the laboratory sample rather than to the environment from which the sample was taken. Clearly, unless equal care is taken with the sampling and pre-analysis treatment processes, the final instrumental data may be useless. In this vein I can do no better than to quote from the Preface by L. L. Ciaccio in a recently issued pollution handbook: "To obtain knowledge of the specific species present and of their concentrations, the effect of the analytical treatment should in no way change the system with its unique equilibrium and components. The analytical results should characterize the original system and not one that is a modification created by the analytical processing."

Consideration of one method of analysis--atomic spectrometry--to the exclusion of all others surely requires no apology. The incredible growth of atomic absorption spectrometry over the past decade bears witness to the popularity and widespread utility of these techniques, even allowing for the suspicion that frequently the equipment is installed for quite the wrong reasons. Atomic spectrometry is ideally suited to solution analysis, so that it is curious that it has been used mostly in fields other than natural water analysis, and that no current text emphasizes this type of analysis. I hope that this contribution will go some way toward remedying this omission. It has been my intention also to consider the application of atomic spectrometry to this specific field in a somewhat different fashion from that employed by my predecessors. The element-by-element "cookbook" approach is, I feel, essentially useless for analysis at the trace level. Where chemical manipulations of the sample are the rule, and where instrumentation is being operated at the limits of performance, a thorough grounding in the underlying principles is a *sine qua non*. Conversely it is not the role of the advanced treatises to emphasize the subordinate role of the analysis methods as but a tool to aid in characterizing the behavior of, in this case, the natural

aqueous systems. I have attempted to chart a middle
course and trust that in some measure this venture
has been successful, else there would be little
excuse for adding to the proliferation of literature
on this topic.

Consideration of the atomic absorption mode of
analysis of necessity dominates the practical discussions, but the interrelationship of absorption,
fluorescence and emission is constantly emphasized
since the potential of these methods for certain
applications is often unappreciated by the working
analyst. Similarly, flame atomization is dealt with
in some detail as the standard choice in most
laboratories, but I have given equal weight to--have
indeed advocated wherever possible--the various
methods of nonflame atomization. These latter devices are revolutionizing the determination of metals
present in trace and ultra-trace quantities and will
without doubt very rapidly dominate the field as far
as natural water analysis is concerned. Much of this
atomization equipment reported in the literature has
been custom-built but several excellent units are
now widely available on the market.

My debt to the many analytical chemists
specializing in atomic spectrometry as well as to
co-workers in the environmental field should be
obvious from the cited references. I wish to grate-
fully acknowledge the help over the years of my
associates in the Institute of Marine Science of the
University of Alaska, and particularly the help of
Ray Hadley, Meng-Lein Lee, Valerie Williamson and
Gulser Wood. I also wish to acknowledge the con-
tinuing financial assistance from the U.S. Atomic
Energy Commission and the National Science Founda-
tion. I am most appreciative of the efforts of Rick
Grunder, Marti Johnson and Lavonia Wiele who typed
the manuscript and of Judith Henshaw and Shirley
Wilson for redrafting the figures.

Fairbanks, Alaska   David C. Burrell
February 1974

CONTENTS

CHAPTER 1

NATURAL WATER SYSTEMS: AN INTRODUCTION

"The actual natural water systems usually consist of
numerous mineral assemblages and often of a gas phase in
addition to the aqueous phase; they nearly always include
a portion of the biosphere. Hence natural aquatic habi-
tats are characterized by a complexity seldom encountered
in the laboratory."

W. Stumm and J. J. Morgan[1]

Before attempting to formulate a sampling and
analysis scheme for aqueous trace heavy metals, it is
clearly necessary to consider, at least in broad outline, just what chemical and physical forms are
likely to occur. Because these systems are not static
as are, for example, rock systems, it is important to
understand as far as possible the pathways followed
between the various environments, and the dynamic
relationships of the constituent components of each.
   It is not possible, of course, to consider the
physical chemistry of natural waters in the same way
that simple laboratory solutions may be studied. The
former is exceedingly complex, involving a gas and
various solid abiotic phases in addition, usually,
to parts of the biosphere. Reaction rates may be
extremely slow and metastable phases abound in the
real world. In spite of this, however, considerable
success has attended efforts to thermodynamically model
the more abundant components of judiciously selected
subsections of the hydrosphere so that, in very general
terms, it is possible to predict the "natural order"
major component chemistry of many environments.
   The same cannot, unfortunately, be said at the
present time for minor constituents such as the
heavy-metal populations. Although impressive progress

1

has been made in this direction also, a great deal remains to be learned of the character and pathways followed by these elements under most natural conditions, and it is often difficult to predict the effect of additional heavy-metal "stress," *i.e.*, pollution.* The heavy metals in solution are highly reactive--hence their trace level concentrations--and may be removed by a variety of physicochemical and biochemical processes to the biotic and inorganic particulate phases. Although these reactions grossly complicate attempts to model and predict distributions on a fundamental chemical basis, they provide considerable buffering capacity for the water.

This introductory chapter can summarize a few of these topics in a most cursory and basic fashion only, and the reader will undoubtedly feel the need for frequent recourse to the cited literature.

THERMODYNAMIC MODELS

The results of efforts to theoretically model various types of natural water systems have been reviewed in detail by, for example, Stumm and Morgan[1] and Schindler.[2] The incentive for this type of work is obviously very great since it offers the promise of a comprehensive characterization of specific environments and insight into the probable effects of imposing chemical and physical stresses and constraints.

Real water systems are "open" in that both energy and matter may be exchanged with their surroundings. Formulation of a thermodynamic equilibrium model, therefore, initially requires designation of some idealized closed portion in which exchange of matter across the boundaries may be reasonably neglected; the equilibrium (time invariant) state of a closed system is that of maximum stability. Once the phases to be included in the model have been specified, and the correct components identified, equilibrium compositions may be computed using free energy and equilibrium constant data.[3] Several such studies of freshwater systems have been published,[4] but possibly the best known example of an equilibrium model construction

---

*The term "pollution" is used in this volume in the broadest sense as a nonnatural addition to a natural water system; it does not necessarily imply demonstrable harm to man or to any trophic level.

is that devised for seawater by Sillén.[5] Of all the world's waters, the open oceanic areas might be expected to most closely conform to the conditions required for a reasonable correspondence between fact and theory, and data determined for this system are obviously of more than local interest. Several other seawater models[6] have been produced following Sillén's pioneering work. The high ionic strength of marine water, however, makes it something of a difficult medium to work with since most available thermodynamic constants relate to an ideal activity scale. The nonideality of natural waters--*i.e.*, the problem of relating concentrations to activities--is a constant difficulty in this field.[7]

The utility of these mathematical manipulations, of course, depends upon the degree to which the calculations are able to predict the concentration and chemical characteristics of actual bodies of water. As far as the minor heavy-metal constituents of seawater (for which the most data exist) are concerned, the results generally demonstrate glaring discrepancies (see Reference 8 and Table 10.4); however, the values calculated for cadmium, copper and zinc by Schindler[9] are less discouraging. Very briefly, the more important factors likely to account for the lack of correspondence between calculated and observed distributions may be summarized as follows:

1. Poor or incomplete choice of the pertinent reactions and participating chemical species. *All* the equilibria solution species must be considered, but there are considerable gaps in our information in this area. (A simple illustration involving silver complexes is included in Chapter 10.)
2. The natural rates of certain thermodynamically predicted reactions may be extremely slow.[10] Methyl mercury is one example of an apparently metastable heavy-metal complex of some importance.
3. Analysis data may be inadequate or insufficiently precise to test the model. Even if determined total quantities are approximately correct, the partition between phases and species divisions is frequently not known. This is why procurement of the latter type of information is stressed in the following pages.
4. The metals may be removed from solution by processes--incorporation in, or sorption onto various solid phases as discussed below--not readily amenable to theoretical treatment by these methods.
5. The equilibrium model chosen may not be suitable to describe the particular environment.

The latter objection has been reviewed in detail by Stumm and Morgan.[1] Clearly no natural water system is closed; the degree to which this condition may be presumed to apply depends upon the specific subarea of the hydrosphere isolated for study. Very basically, the change in concentration of some metal species R (Appendix 1 lists the symbols and abbreviations used in this book) in a body of water may be described:

$$\frac{d[R]}{dt} = f\ (x,y,z) \qquad (1.1)$$

where x, y, z are addition/removal, advection, and diffusion or turbulent transport terms. In the case of the oceans, it would appear that the major constituents have remained constant, at least during recent geological time; *i.e.*, the rate of input has been balanced by the removal (sedimentation) rate so that, for these major species, $d[R]/dt \sim 0$. This characteristic conforms to the time-invariant (steady state) condition of an *open* thermodynamic system. The residence time (T) of an element in the oceans has been defined:[11]

$$T = \frac{[R]}{d[R]/dt} \qquad (1.2)$$

and it has been shown by Morgan[12] that when the reaction rates of a constituent are rapid in comparison with the residence time (*i.e.*, $T \gg T_{1/2}$ where $T_{1/2} = \ln2/k$ = the half-time of a first order reaction with rate constant k), then the stationary steady state condition approaches the equilibrium state of the closed model. Unfortunately, these conditions are not fulfilled by the highly reactive, short residence time of heavy metals, although it would appear that the distribution of, for example, the major alkali and alkaline earth elements in large bodies of water might be considered in terms of an equilibrium model if other necessary conditions were met.

The residence time concept can be applied also to smaller, freshwater bodies, but in these cases it is useful to think in terms relative to the residence time (T_w) of the water itself:

$$T_{rel} = \frac{[R]}{d[R]/dt_{input}} \cdot \frac{1}{T_w} \qquad (1.3)$$

This modification provides for the reactions with coexisting solid phases which are ubiquitous with the heavy metals. For example, T_{rel} will be less than or will exceed unity in cases where the metal is removed from solution by uptake or sorption, or is remobilized.

SOLUBLE HEAVY-METAL SPECIES

The natural concentrations of heavy metal are extremely low in the hydrosphere (see also Chapter 2) and it might initially appear to be reasonable to consider permissible insoluble species as the primary controlling mechanism. In most environments, oxide, hydroxide and carbonate solid phases (see Figure 10.5) are of major importance. Figure 1.1[1] shows the solubility of various metals as a function of pH based on ideal solubility equilibria of the type:

$$R(OH)_n \rightleftharpoons R^{n+} + nOH^- ; \qquad K_{sp} = [R^{n+}][OH^-]^n \qquad (1.4)$$

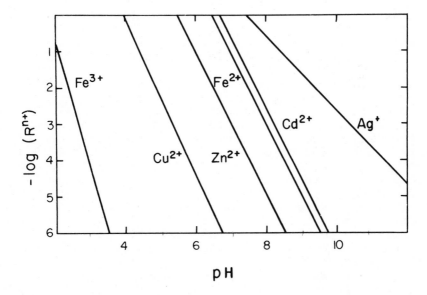

Figure 1.1. *Concentrations of aquo ions in equilibrium with solid (hydr)oxides. (Data from Reference 1.)*

The utility of this type of information is strictly limited however, since not only must the formation of other soluble complexes in addition to or in place of the basic aquo ion be considered, but sorption and uptake depletions to the solid phases are of major importance. It is primarily for these latter reasons that most trace-metal concentration values determined for natural waters (Reference 8 and Table 10.4) fall well below theoretical predictions based upon least-soluble compound equilibria.

The bare-metal ion included in Equation 1.4 cannot exist in water. The basic unit is the hydrated aquo ion, which is frequently of indeterminate structure.[13] The aquo ion functions as a (Brønsted) acid and, by a series of proton transfer (hydrolysis) reactions, a sequence of hydroxo and possibly oxo complexes may be produced. However, since the pH range of natural waters is quite limited, this reaction series is poorly developed in nature. The relative acidity of the aquo ion, and hence the nature of the coordinate complexes stable over the natural water pH range (see Figure 1.3), is a function of the ionic potential (charge/radius ratio) as illustrated in Figure 1.2 (see also Reference 14). As might be supposed, given the ubiquitous occurrence of hydroxyl ions, hydrolysis reactions are of considerable importance in these waters. Representative hydrolysis complexes are listed in Table 1.1. Stumm[1,15] has considered this topic in some detail and has discussed the formation of polynuclear complexes leading to hydrous oxide precipitation.

Exchange reactions with other common unidentate ligands in freshwater media are unimportant. In seawater, however, although hydroxo complexes probably predominate, coordination reactions with other major inorganic ligands are important for the trace-metal constituents. The metals of Table 10.4, for example, are believed to form stable chloro complexes. The thermodynamic conditions necessary for the predominance of halogeno over hydroxo complexing have been considered elsewhere.[5,17] Some soluble heavy-metal species believed to be stable in marine waters are tabulated in Table 1.2, but it should be noted that considerable work remains to be done in this field.

In contrast to the common inorganic unidentate ligands, organic chelating ligands would theoretically be expected to form very stable soluble complexes with a range of heavy metals even when these groups are present only in trace amounts. There is, in

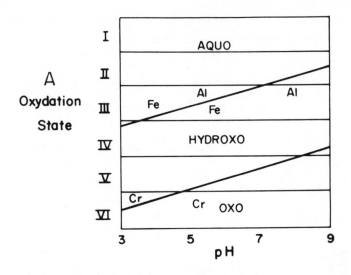

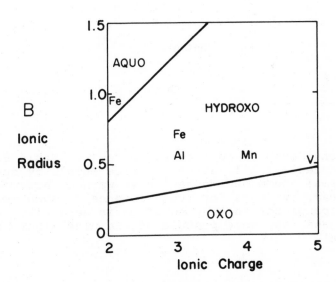

Figure 1.2. Soluble metal complex in water.
A. Oxidation state *vs.* pH. (Data from References 1 and 15.)
B. Ionic charge *vs.* radius at environmental pH range. (After Reference 16.)

Table 1.1

Some Examples of Hydrolysis Complexes
(Data from Reference 15)

Ag (I)	$AgOH$, Ag_2OH^+, $Ag_2(OH)_2$, $Ag(O)_2^-$
Bi (III)	$BiOH^{2+}$, $Bi_6(OH)_{12}^{6+}$, $Bi_9(OH)_{20}^{7+}$, $Bi_9(OH)_{22}^{5+}$
Cr (III)	$CrOH^{2+}$, $Cr(OH)_2^{4+}$, $Cr_6(OH)_{12}^{6+}$, $Cr(OH)_4^-$
Cu (II)	$CuOH^+$, $Cu_2(OH)_2^{2+}$, $Cu(OH)_3^-$, $Cu(OH)_4^{2-}$
Fe (III)	$FeOH^{2+}$, $Fe_2(OH)_2^{4+}$, $Fe(OH)_2^+$, $Fe(OH)_4^-$
Hg (II)	$HgOH^+$, Hg_2OH^{3+}, $Hg_2(OH)_2^{2+}$
In (II)	$InOH^+$, In_2OH^{3+}, $In(OH)_3^-$
Mn (II)	$MnOH^+$, $Mn(OH)_3^-$
Pb (II)	$Pb_4(OH)_4^{4+}$, $Pb_6(OH)_8^{4+}$, $Pb_3(OH)_4^{2+}$, $Pb(OH)_3^-$

fact, considerable indirect evidence (see Table 2.2) for enhanced solubilization of trace metals by such mechanisms; Stumm and Morgan[1] have warned, however, that *direct* evidence for the existence of such multidentate complexes in true solution is generally lacking and that some postulated examples might be colloidal.

REDOX EQUILIBRIA

Heavy-metal species in solution strive for improved stability in another way through reactions involving oxidation state changes, *i.e.*, by electron transfers. Such redox reactions--in which the oxidants and

Table 1.2

Suggested Dissolved Inorganic Species of Heavy Metals in Seawater
(Data from References 18-20 except as noted)

Ag	$AgCl_2^-$; $AgCl_3^{2-}$
As	$HAsO_4^{2-}$ [a]; $H_2AsO_4^-$
Au	$AuCl_2^-$ [a]; $AuCl_4^-$
Bi	BiO^+ [b]; $BiOCl$ [b]; $BiCl_4^-$ [b]
Cd	Cd^{2+}; $CdCl_2^0$ [c]; $CdCl^+$ [d]
Co	Co^{2+}; $CoCl^+$; $(CoSO_4^0)$
Cr	CrO_4^{2-} [b]; (OH species)
Cu	Cu^{2+}; $CuCO_3^0$ [c]; $(CuSO_4^0)$; (OH, Cl species)
Fe	$Fe(OH)_2^+$ [c]; $Fe(OH)_4^-$ [c]
Hg	$HgCl_3^-$; $HgCl_4^{2-}$; $HgCl_2^0$
Mn	Mn^+; $MnSO^0$; (OH species) [a]
Mo	MoO_4^{2-}
Ni	Ni^{2+}; $NiSO_4^0$; (Cl species)
Pb	Pb^{2+}; $PbSO_4^0$; $PbCl_3^-$; (OH species) [a]
Sb	$Sb(OH)_6^-$ [a]
Sn	(OH species) [a]
Ti	$Ti(OH)_4$? [a]
Tl	Tl^+
V	$VO_2(OH)_3^{2-}$; $H_2VO_4^-$ [b]; $H_3V_2O_7^-$ [b]
W	WO_4^{2-}
Zn	Zn^{2+}; $ZnCl^+$ [b]; $ZnSO_4^0$ [c]; $Zn(OH)^+$ [d]; $ZnCO_3^0$ [e]; $Zn(OH)_2^0$ [e]

a, b, c, d, e Data from References 5, 8, 9, 21, and 22
respectively.

reductants are electron recipients and donors
respectively--are conceptually analogous to acid-base
(proton flux) reactions, so that a relative electron
activity scale may be defined such that $p\varepsilon = - \log a_\varepsilon$,[2]
which is, for a given temperature, a direct function
of the redox potential (Eh). It is possible, in theory
to relate measured Eh potentials to the solution
composition *via* the Nernst equation:

$$Eh = E° + \frac{2.3RT}{nF} \log \frac{(oxidant)}{(reductant)} \qquad (1.5)$$

where n gives the number of electrons transferred and
E° is the standard electrode potential.[1,24] Since
(balanced) aqueous reactions will include H or OH
within the activity quotient of Equation 1.5, it is
possible--and common practice--to represent ideal
equilibria of selected aqueous systems by Eh-pH
stability-field diagrams as illustrated by Figures
10.1 and 10.6.

It is important to appreciate the practical
limitations of these constructs. In the first
place, it is obviously impossible to represent much
information by means of a simple binary diagram;
the chief casualty here is knowledge of the reactant
concentration dependence. This is, however, a problem
of representation only since, in theory, there is no
limit to the number of figures which could be constructed to represent any given system. The major
reservations are the same as those for the thermodynamic models considered above; namely, the need to
identify *all* the possible reacting species, and the
need to develop a reasonable approach when dealing
with discrepancies of fact and theory with regard to
a state of thermodynamic equilibrium. The latter
reservation is particularly damning in the case of
redox models since most natural water reactions are
exceedingly slow and only partial equilibria are to
be expected. It has been noted[25] in fact that, at
the redox potential of normal, aerated surface waters,
virtually all elements should exist in their highest
oxidation states. Since this is not the case, it is
clear that true equilibrium systems are rare in
nature and that metastability is the rule rather than
the exception. At best, Eh-pH constructions can
indicate the *direction* of the natural reactions, and
field boundaries should serve as a guide only to the
expected existence of particular chemical species.
Figure 1.3 shows--very approximately--the boundaries
of the Eh-pH environments of surface waters. The

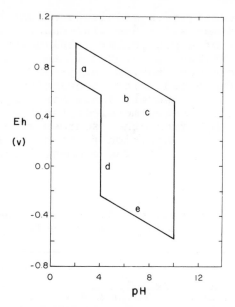

Figure 1.3. *Approximate natural surface water Eh-pH field boundaries. (After Reference 24.)*

[a] *Acid mine drainage*

[b] *Normal lakes and rivers*

[c] *Normal marine water*

[d] *Bog waters*

[e] *Anoxic environments.*

practical difficulties of measuring environmental redox potentials have been exhaustively considered in the standard literature.[1]

UPTAKE BY ORGANISMS

Aquatic organisms are capable of accumulating quantities of heavy metals many orders of magnitude in excess of the coexisting soluble contents.[26-28] There is no general consensus as to the effects of these processes on aqueous concentrations but they are generally thought--except possibly for deep oceanic areas--to be subordinate to the control exercised by sorption onto inorganic particulates. In a few coastal areas, seagrass stands may form localized reservoirs of a magnitude comparable to that of the sediments.[29] See also Chapter 2.

The quantity of a particular metal taken up by an organism is frequently quoted in terms of a concentration factor over the amount present in the water[30] as in Table 1.3. This convention is frequently misleading, however, since many organisms do not derive their heavy-metal contents directly from the water, but from other sources (food, sediment); thus "concentration factors are indicators only and not absolute."[31] The elemental concentration

Table 1.3

Accumulation and Transfer of Selected Trace Metals
in Plankton (Data from Reference 31)

| Element | Concentration[a] | | Conversion[b] (%) Phytoplankton to Zooplankton |
	Phytoplankton	Zooplankton	
Ag	23,000	9,000	25
Co	1,500	700	30
Cr	4,400	1,900	28
Cu	30,000	6,000	13
Fe	45,000	25,000	40
Ga	8,000	7,000	60
Mn	4,000	1,500	20
Ni	5,000	3,000	39
Pb	40,000	3,000	5
Sn	6,000	450	5
Ti	25,000	6,800	17
Zn	26,000	8,000	20
Zr	60,000	25,000	30

[a]Ratio of concentration in organisms over water; see text.

[b]Conversion efficiency of Lowman *et al.*[31] based on 50% C conversion and assuming the metal to be derived from food.

obviously represents a balance between uptake and loss, and the factors controlling the rates of the relevant processes for the heavy metals are not known in most cases. Much of the available information relates to radionuclides[28],[31] and, as noted by Wolfe and Rice,[32] such data may not necessarily be representative of the stable metal reactions or contents. In many aquatic systems the plankton constitute the largest biotic reservoir for the heavy metals; the concentration factors given for the phytoplankton in Table 1.3 should represent a valid comparison with the soluble aqueous values. Higher trophic level populations make up a considerably smaller proportion of the total biomass, but the heavy metals present in species harvested as food sources are obviously of enhanced interest.

It is frequently convenient to consider the relationship of the heavy metals to organisms in terms of required and inessential (including toxic) constituents. The former group--basically cobalt, copper, iron, manganese and zinc--functions in various physiological roles (enzyme activation, redox and transport processes, etc.). Of course, the same elements are not necessarily essential to all aquatic species and even these metals may be toxic if present in high concentrations. Generally, enrichment factors conform to the order of stability of complexes formed with a number of basic organic ligands.[20],[26],[33]

It is important to fully appreciate that metals may be extracted from solution by nonspecific physical adsorption processes or *via* the ingestion of particulate matter, and to realize that amounts obtained in this fashion may greatly exceed the physiologically bound fraction. Sorbed metals may be subsequently assimilated into the organism. Benthic biota appear to be major concentrators of a wide range of heavy metals, and this is hardly surprising in view of the enhanced available quantities which are associated with the sediments as compared with those quantities associated with the overlying waters. Sorption onto surface membranes has been cited as being of major importance in many individual cases,[34],[35] but has not been systematically studied in any detail. These processes are particularly important with regard to the toxicity effects of heavy metals on fish, which have been shown[36] to be primarily attributable to the formation of mucous-metal precipitates on the gills. Such a multiplicity of uptake mechanisms severely complicates consideration of food web

transfer. Table 1.3 includes estimates of phytoplankton to zooplankton conversion factors (based
upon a 50% carbon conversion[31]). These calculations
assume, of course, that the zooplankton receive their
heavy metals primarily from food. No unequivocal
example of a food chain *magnification* of a toxic
metal is known to the author.

ABIOTIC SORPTION

All solid particles in natural waters have electrically charged surfaces, and, given the large
surface areas of normal particulate matter in these
waters (up to $10-10^2$ m^2/g for clay minerals), the
resultant surface charge density (say 10^4-10^5 esu/cm^2)[3*]
will clearly exhibit a profound effect on the distribution of heavy-metal complexes. Such interfacial
phenomena may be generally characterized on the basis
of the Gouy electrical double-layer theory,[1] but
this model has limited practical utility since it
can take no account of the relative *sizes* of the
various solute species.

The structure of clay minerals[38] results in
"plate" surfaces and edges with excess negative and
positive charges respectively but with an overall
net negative charge,[39] so that uptake of cationic
species onto this material is of prime importance.
The removal of an ion from solution (at constant T
and pH) may generally be described by means of
various empirical expressions of which the Freundlich
isotherm is probably the most commonly used:

$$y = Const_1 \cdot x^{\frac{1}{Const_2}} \qquad (1.6)$$

At equilibrium y and x represent, respectively, the
quantities of solute species adsorbed per unit mass
of the solid phase and that quantity remaining in
solution.

For many environmental applications, the major
consideration tends to be the uptake selectivity for
particular ions onto the solids. Ion exchange may
be expressed by pseudo mass-action equilibria expressions (compare Equations 5.12 and 5.13), but the
actual replaceability depends upon a host of factors
peculiar to the specific ion, and to the physical
and chemical environment of both the solution and

the solid surface.[40] Several replacement series
have been proposed for the alkali and alkaline earth
elements which are related in a rough fashion to
ionic charge and size. Extrapolation of these con-
cepts--for the most part empirically derived for
soils--to natural water systems is not easily
accomplished, although the basic relationships
appear to hold for the major constituents; magnesium,
for example, rather than the monovalent ions, is
preferentially (and rapidly)[41] taken up on clays
injected into seawater.[42] There is ample evidence
to indicate that many trace metals are very efficiently scavenged from freshwater environments by
the suspended solids,[43,44] but selectivity data and
detailed knowledge of behavior under varying physical
conditions are sparse.

This brief discussion has so far emphasized the
role of clay minerals. However, other solid phases
are important in these waters either as primary
constituents or as surface impurities on clay
minerals. Of these, probably the most important are
the various hydrous oxides[45,46] whose surface charge
is a function of pH. The negative charge characteristics of the iron and manganese oxides increase
with pH so that, for example, it might be expected
that an increased uptake of cations would accompany
passage of this material across the fresh-marine
interface. O'Connor and Renn[47] have, in fact,
demonstrated increased sorption of zinc onto natural
river particulates with pH, but other experimental
evidence (see the following chapter) appears to
argue for limited *desorption* of heavy-metal species
in marine water. Pravdić's study of electrokinetic
potentials[48] has shown that, although sediments
suspended in freshwater are invariably negatively
charged, a charge reversal commonly occurs in seawater. His work would thus also appear, *a priori*,
to support the concept of cationic release in seawater. One of the principal problems encountered
in attempting to resolve these and similar difficulties is--as discussed previously--the large measure
of ignorance as to the nature of the soluble heavy-
metal species. Taking seawater again as an example,
it may be seen from Table 1.2 that a number of the
more common heavy metals have been postulated to
exist in this medium as stable anionic complexes,
and that the "cationic complex" metals are frequently
those for which some evidence exists for desorption
or, at least, for which uptake onto sediments is not
a marked characteristic (*e.g.*, cadmium; see Chapter
10). It would be expected that aquo ions should be

less strongly bound to solid surfaces than metals coupled to other inorganic ligands; Stumm and Morgan[49] have noted that hydrolysis species are particularly strongly adsorbed. Lowman *et al.*[31] have emphasized the probable importance of the association of epiphyton and particulate material in trace-metal sorption processes. There is also evidence for the removal of heavy-metal ions from solution by nonsorption processes. Ahmed,[50] for example, has demonstrated the direct *precipitation* of a nickel nitrate-hydroxide onto various substrates.

REFERENCES

1. Stumm, W., and J. J. Morgan. *Aquatic Chemistry*, John Wiley and Sons, New York, 1970.
2. Schindler, P. W. in *Impingement of Man on the Oceans*, (D. W. Hood, ed.), John Wiley and Sons, New York, 1971, pp. 219-244.
3. Denbigh, K. *The Principles of Chemical Equilibrium*, Cambridge University Press, Cambridge, 1961.
4. Morel, F., and J. Morgan. *Environ. Sci. Technol.*, *6*, 58 (1972).
5. Sillén, L. G. in *Oceanography*, (Mary Sears, ed.), Amer. Assoc. Adv. Sci., Publ. No. 67, Washington, D.C., 1961, pp. 549-582.
6. Kramer, J. R. *Geochim. Cosmochim. Acta*, *29*, 921 (1965).
7. Sillén, L. G. in *Equilibrium Concepts in Natural Water Systems*, (R. F. Gould, ed.), Advan. Chem. Ser. No. 67, American Chemical Society, Washington, D.C., 1967, pp. 45-56.
8. Krauskopf, K. B. *Geochim. Cosmochim. Acta*, *9*, 1 (1956).
9. Schindler, P. W. in *Equilibrium Concepts in Natural Water Systems*, (R. F. Gould, ed.), Advan. Chem. Ser. No. 67, American Chemical Society, Washington, D.C., 1967, pp. 197-221.
10. Garrels, R. M. in *Researches in Geochemistry*, (P. H. Abelson, ed.), John Wiley and Sons, New York, pp. 25-37.
11. Barth, J. F. *Theoretical Petrology*, John Wiley and Sons, New York, 1952.
12. Morgan, J. J. in *Equilibrium Concepts in Natural Water Systems*, (R. F. Gould, ed.), Advan. Chem. Ser. No. 67, American Chemical Society, Washington, D.C., 1967, pp. 183-195.
13. Horne, R. A. *Marine Chemistry*, John Wiley and Sons, New York, 1969.
14. Tyree, S. W. in *Equilibrium Concepts in Natural Water Systems*, (R. F. Gould, ed.), Advan. Chem. Ser. No. 67, American Chemical Society, Washington, D.C., 1967, pp. 183-195.

15. Stumm, W. in *Principles and Applications of Water Chemistry,* (S. D. Faust and J. V. Hunter, eds.), John Wiley and Sons, New York, 1967, pp. 520-560.
16. Mason, B. *Principles of Geochemistry,* 2nd Ed., John Wiley and Sons, New York, 1958.
17. Martin, D. F. in *Equilibrium Concepts in Natural Water Systems,* (R. F. Gould, ed.), Advan. Chem. Ser. No. 67, American Chemical Society, Washington, D.C., 1967, pp. 255-269.
18. Goldberg, E. D., and G. O. S. Arrhenius. *Geochim. Cosmochim. Acta, 13,* 153 (1958).
19. Goldberg, E. D. in *The Sea,* Vol. 2, (M. N. Hill, ed.), Interscience Publishers, New York, 1963, pp. 3-25.
20. Goldberg, E. D. in *Chemical Oceanography,* Vol. 1, (J. P. Riley and G. Skirrow, eds.), Academic Press, London, 1965, pp. 163-196.
21. Barić, A., and M. Branica. *J. Polarog. Soc., 13,* 4 (1967).
22. Zirino, A., and M. L. Healy. *Limnol. Oceanog., 15,* 956 (1970).
23. Sillén, L. G. *Chemical Equilibrium in Analytical Chemistry,* Interscience Publishers, New York, 1962.
24. Garrels, R. M., and C. L. Christ. *Solutions, Minerals and Equilibria,* Harper and Row, New York, 1965.
25. Morris, J. C., and W. Stumm. in *Equilibrium Concepts in Natural Water Systems,* (R. F. Gould, ed.), Advan. Chem. Ser. No. 67, American Chemical Society, Washington, D.C., 1967, pp. 270-285.
26. Goldberg, E. D. *Geol. Soc. Am. Mem., 67,* 345 (1957).
27. Bowen, H. J. M. *Trace Elements in Biochemistry,* Academic Press, New York, 1966.
28. Åberg, B., and F. P. Hungate (eds.). *Radioecological Concentration Processes,* Pergamon, Oxford, 1967.
29. Parker, P. *Texas Univ. Pub., Marine Sci., 8,* 75 (1963).
30. Polikarpov, G. G. *Radioecology of Aquatic Organisms,* (Translated from Russian by Scripta Technica), Reinhold, New York, 1966.
31. Lowman, F. G., T. R. Rice and F. A. Richards. in *Radioactivity in the Marine Environment,* Nat. Acad. Sci., Washington, D.C., 1971, pp. 161-199.
32. Wolfe, D. A., and T. R. Rice. *Fish. Bull., 70,* 959 (1972).
33. Schubert, J. in *Chemical Specificity in Biological Interactions,* (F. R. N. Gurd, ed.), Academic Press, New York, 1954, pp. 114-163.
34. Korringa, P. *Quart. Rev. Biol., 27,* 266 (1952).
35. Cushing, C. E., and F. L. Rose. *Limnol. Oceanog., 15,* 762 (1970).
36. Katz, M. in *Water and Water Pollution Handbook,* Vol. 1, (L. L. Ciaccio, ed.), Marcel Dekker, New York, 1971, pp. 297-328.

37. Wayman, C. H. in *Principles and Applications of Water Chemistry*, (S. D. Faust and J. V. Hunter, eds.), John Wiley and Sons, New York, 1967, pp. 127-167.
38. Grim, R. E. *Clay Mineralogy*, McGraw-Hill, New York, 1953.
39. van Olphen, H. *An Introduction to Clay Colloid Chemistry*, John Wiley and Sons, New York, 1963.
40. Carroll, D. *Bull. Geol. Soc. Am.*, *70*, 749 (1959).
41. Malcolm, R. L., and V. C. Kennedy. *Soil Sci. Soc.*, *33*, 247 (1969).
42. Carroll, D., and H. C. Starkey. in *Clays and Clay Minerals*, (A. Swineford, ed.), Pergamon, Oxford, 1960, pp. 80-101.
43. Eichholz, G. G., T. F. Craft, and A. N. Galli. *Geochim. Cosmochim. Acta*, *31*, 737 (1967).
44. Gardner, K., and O. Skulberg. *Intern. J. Air Water Pollution*, *8*, 229 (1964).
45. Jenne, E. A. in *Trace Inorganics in Water*, (R. A. Baker, ed.), American Chemical Society, Washington, D.C., 1968, pp. 337-387.
46. Parks, G. A. in *Equilibrium Concepts in Natural Water Systems*, (R. F. Gould, ed.), Advan. Chem. Ser. No. 67, American Chemical Society, Washington, D.C., 1967, pp. 121-160.
47. O'Connor, J. T., and C. E. Renn. *J. Am. Water Works Assoc.*, *52*, 1055 (1964).
48. Pravdić, V. *Limnol. Oceanog.*, *15*, 762 (1970).
49. Stumm, W., and J. J. Morgan. *J. Am. Water Works Assoc.*, *54*, 971 (1962).
50. Ahmed, A. Ph.D. Dissertation, Department of Mineral Engineering, Stanford University, Palo Alto, Calif., 1971.

CHAPTER 2

AQUEOUS HEAVY-METAL POLLUTION

"The two groups of chemicals that appear to offer
the greatest dangers through promiscuous release to
the environment are the heavy metals and halogenated
hydrocarbons."

National Academy of Science[1]

There is, of course, no exact meaning to the term
"heavy metal" (or to the term "pollution" either, but
this is considered more fully below) and the phrase
has been deliberately selected to imply ambiguity.
It is intended that this book shall cover in a general
fashion the AS analysis of the block of elements
shown in Figure 2.1. This includes metals of current

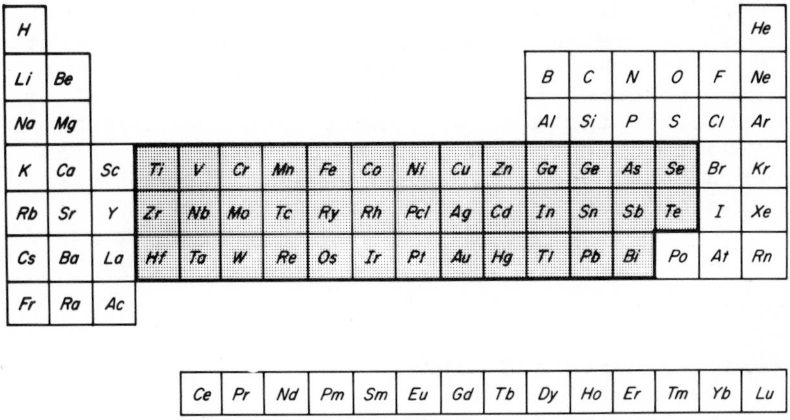

H																	He
Li	Be											B	C	N	O	F	Ne
Na	Mg											Al	Si	P	S	Cl	Ar
K	Ca	Sc	Ti	V	Cr	Mn	Fe	Co	Ni	Cu	Zn	Ga	Ge	As	Se	Br	Kr
Rb	Sr	Y	Zr	Nb	Mo	Tc	Ry	Rh	Pcl	Ag	Cd	In	Sn	Sb	Te	I	Xe
Cs	Ba	La	Hf	Ta	W	Re	Os	Ir	Pt	Au	Hg	Tl	Pb	Bi	Po	At	Rn
Fr	Ra	Ac															

	Ce	Pr	Nd	Pm	Sm	Eu	Gd	Tb	Dy	Ho	Er	Tm	Yb	Lu

Figure 2.1. *Heavy metals as defined in this volume. (Also see Appendix 4.)*

concern as major potential polluters--the first
transition series, silver, arsenic, cadmium, mercury,
lead, selenium, etc.--and also those allied elements
which are not presently widely discharged in major
quantities but which might be expected to behave
similarly. The latter group may also now or in the
future be of localized concern. To some extent, it
might be reasonably expected that the emphasis in
aqueous metal pollution will shift with time as new
industrial processes make use of more "exotic"
metals or, possibly, as the use or disposal of other
metals is legislatively restricted.

Consideration only of the "heavy metal" suite
implies that the primary concern of the techniques
discussed is with "trace metal" analysis and also
implies that the alkali and alkaline earth metals
are excluded. The latter series include the major
cations of all natural waters; the distribution of
the heavier and rarer members of these groups is
controlled by the more common analogs. Several of
these metals are important from the pollution standpoint (beryllium, for example) and would warrant a
separate analytical volume. In contrast to the
metals dealt with in the following pages, an AS
analysis program for the alkalis and alkaline earths
would rely heavily upon direct (flame) emission
techniques.

The reasons for the dispersal of heavy metals in
extremely low concentrations within the hydrosphere
(except in very rare instances) have been briefly
outlined in the previous chapter. No rigid definition
of that much-abused term "trace metal" will be adhered
to, but the reader should infer aqueous concentrations
in the $\mu g/\ell$ range (ppb). Absolute concentrations
will be referred to in conjunction with several of
the specific techniques. These data must not be
allowed to obscure the primary requirements of an
environmental analysis which will be for a *concentration* within a given solution or solid matrix.

DISTRIBUTION

Natural Pathways and Sinks

The elements of Figure 2.1 are highly reactive
and generally widely dispersed in nature in association with or as part of a variety of solid phases.
Residence times in solution in natural waters are
hence very short and these heavy metals constitute

trace constituents within the hydrosphere. As noted
in the previous chapter, it is theoretically possible
to predict from available thermodynamic data the
equilibrium distribution of any specific metal within
a variable pH/Eh field (see Figure 10.1) so long as
the components and physical parameters of the system
are limited and closely defined. The natural pathways
of heavy metals are largely controlled by two major
processes: incorporation on or into various solid
phases, and chemical complexation in solution.
These mechanisms act to restrict and promote solubilization of the metal respectively, but immobilization
to the various solid "sinks" predominates, and
trace-metal concentrations in all natural waters
tend to be less than might be calculated (see Table
10.4, for example) from a reasonable geochemical
knowledge of the environment. The kinetics of these
reactions also plays a major and usually poorly
understood role.

It is not considered to be wise at this time to
cite "mean" values for the heavy-metal contents of
either fresh or marine waters since reliable analytical
techniques have only recently become widely available
and since it is still usually impossible to determine
which fraction of the aqueous content has been included in many of the published compilations. In
the case of freshwater bodies, average content data
have little general utility. The offshore marine
waters are unlikely to vary beyond narrow limits over
wide areas of the oceans but, mostly because of
analytical difficulties, very few reliable data are
as yet available and many of the extant values are
biased in favor of coastal areas. The oft-cited
numbers given by Goldberg[2,3] may serve as an order-
of-magnitude guide for most of the common heavy
metals, but advances in analytical technique in the
intervening decade have rendered some of these values
redundant or doubtful (see Table 5.1). As a guide
to the performance required of any analytical system,
Table 2.1 lists river water values for two arbitrary
U.S. river systems in well-populated areas (see also
Figure 2.2); a few data for cadmium, copper, lead
and zinc for unpolluted lake water are given in
Table 10.5.

A considerable fraction of the soluble (or
colloidal) heavy-metal content of freshwater bodies
may be in the form of a wide range of organic complexes. Depending upon the choice of processing and
analysis technique, this fraction may potentially be
overlooked in whole or in part by the analyst.
Table 2.2, for example, lists copper and lead data

Table 2.1

Some Heavy-Metal Concentrations in Two U.S. River Systems
(μg/ℓ)

	California Rivers (Means or Ranges) (Reference 5)	Maine Rivers (Means) (Reference 4)
Bi	0.7	
Cd	n.d.	
Co	0.7–15	
Cr	n.d.	0.3
Cu	18	12
Fe	33	
Ga	n.d.	
Mn	7	4
Mo	1–5	
Ni	1–5	0.3
Pb	6	2
Ti	7	3
V	3	
Zn	30	

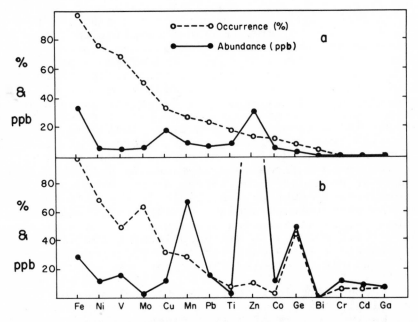

Figure 2.2.   *Concentrations (μg/ℓ) and percentage occurrence of*
certain heavy metals in natural waters of California.
(Data from Reference 5 by courtesy of author.)

[a]*65 stream samples*   [b]*72 spring-water samples*

Table 2.2

Cu and Pb Values for Certain U.S. Polluted River and Lake Waters[a] (μg/ℓ)   *(Data from Reference 6)*

Location	Sample	Cu		Pb	
		Free[b]	Acid[c]	Free[b]	Acid[c]
Rouge River	1	2.0	27.0	1.1	15.0
	2	14.0	108.0	2.2	37.0
	3	8.0	19.0	0.4	11.0
Detroit River	4	18.0	28.0	0.4	5.0
	5	11.0	29.0	0.4	6.0
Lake Erie	6	0.5	1.8	0.1	1.6
Lake Michigan	7	0.8	19.0	0.2	3.3

Samples: 1 - half-mile above Ford Co. turning basin; 2 - from turning basin; 3-- river mouth; 4 - Trenton channel; 5 - Livingston channel; 6 - surface central basin; 7 - at Waukegan, Illinois

[a] Analysis by anodic stripping voltammetry

[b] Untreated sample

[c] Sample acid treated; see text.

(by anodic stripping voltammetry[6]) for untreated polluted lake and river water samples and for samples treated with acid to release the sequestered fraction. The high molecular weight humic-acid contents of a range of waters is currently the subject of extensive research,[7] and it has been suggested[8] that this material may play a major role in mobilizing and transporting heavy metals. In marine water, heavy metals exist predominantly as complexes with the major inorganic ligands. The species listed in Tables 1.2 and 10.4 have, for the most part, been predicted from limited thermodynamic data, and confirmation of the existence of these compounds is urgently required so that solubilities and uptake mechanisms within specific environments may be more closely predicted.

The heavy trace metals are removed from solution by uptake onto or incorporation into various biotic and abiotic solid phases. But this immobilization may be only relative since the solid "sinks"--inorganic particulates or fish species, for example--are themselves liable to passive or active transportation.

Remobilization of the metals from the solids may
occur with changes in the prevailing physical or
chemical environment, *e.g.*, through burial of the
sediment or death of the organism. As noted in the
previous chapter, there is currently considerable
speculation concerning the behavior of metals sorbed
to suspended sediment during passage across the
fresh-marine interface. The results of some laboratory experiments[9] appear to argue for desorption of
some metals (*e.g.*, silver, cobalt, zinc) at this
boundary as a function of pH, concentration and
other physical parameters. These and similar experiments have invariably utilized radio-tracers, and
it has been shown[10] that exchange between such
tracers and the various naturally occurring metallic
species is not necessarily easily or rapidly
accomplished. If, in fact, desorption of the heavy
metals at the marine margins were to be the rule, the
concentrations of these metals in seawater would be
expected to more closely approach the values thermodynamically predicted. At the present time the
majority of evidence points to the estuarine and
coastal deposition of river-borne sediment as being
the major sink within the heavy-metal cycle, and one
of the major deficiencies in our knowledge of the
environmental distribution of these metals concerns
rates and reactions of potential remobilization
mechanisms from these phases. As just one example,
Windom[11] has demonstrated the important role of
salt-marsh macrophytes as a natural conveyor for
transporting mercury from the sediments back into
the aqueous environment in a more biologically
active form (see also Reference 12).

Pollutional Stresses and
Environmental Protection

    The term "pollution" may be defined in a variety
of ways depending upon one's individual viewpoint.
The ardent conservationist might deplore *any* interference with a natural waterway. This view would
not, however, be shared by the populace at large
since the common conception of pollution[13] is an
aesthetic one, concerning visual and sensory
attributes--"murkiness," for example.
    From a scientific standpoint we may, following
Stumm,[14] define three categories of pollutional
perturbation:

1. direct toxicity
2. interference with the respiration-photosynthesis balance
3. impairment of the *diversity* and/or stability of an ecosystem.

The best known and most publicized cases of heavy-metal pollution have been related to category (1)--mercury and cadmium poisoning in Japanese waters[15],[16] for example. But, for reasons discussed more fully elsewhere, this type of impingement--although dramatic--tends to be localized and relatively easily remedied or forestalled. Our principal concern should be with the latter two divisions above and, regarding specifically the effects of heavy metals, we are on very uncertain ground. So little is known of the "natural" distributions and pathways that prediction of the effects of man-induced perturbations--on reaction rates, on the natural balance of the ecosystems--is exceedingly difficult.

The industrial sources of the major heavy metals likely to be discharged into the hydrosphere are indicated in Table 2.3, although this type of information is in a constant state of flux as new industrial processes are developed. The industrial consumption

Table 2.3

General Distribution of Heavy Metals in Particular Industrial Effluents

	Ag	As	Cd	Cr	Cu	Fe	Hg	Mn	Ni	Pb	Se	Ti	Zn
General Industrial and Mining				X	X	X		X		X			X
Plating			X	X	X				X	X			X
Paint Products			X							X		X	
Fertilizers		X	X	X	X	X	X	X	X				X
Insecticides/Pesticides	X			X		X							
Tanning	X			X									
Paper Products				X	X		X		X	X		X	X
Photographic	X			X									
Fibers					X								X
Printing/Dyeing				X						X			
Electronics	X										X		
Cooling Water				X									
Pipe Corrosion						X				X			

of beryllium (a metal not specifically considered in this book), for example, has increased over 500% in the last few years. With regard to the more immediate effects on man (in the broadest sense) one might consider impingements within the following categories:

1. major health hazards, poisonous and cumulative; *e.g.*, lead and mercury
2. not harmful to man in "small" doses but toxic to lower organisms including those used as food; *e.g.*, copper and zinc
3. discharges which are predominantly aesthetically objectionable; *e.g.*, iron, manganese and titanium.

Silver, arsenic, cadmium, mercury, lead and selenium are considered in some detail in Chapter 10. The following briefly summarizes the potential impact of the remaining metals of Table 2.3.

Chromium. Frequently present (as Cr VI) in industrial wastewaters; commonly used as a rust inhibitor. Poisonous but noncumulative.

Copper. Derived from many industrial processes; important product of boiler corrosion. Toxic to fish, algae, bacteria; hence frequently used to control excessive algal blooms (eutrophication).

Iron. Objectionable primarily because of taste and color in water and staining/deposition characteristics. Common derivatives of general corrosion, but the worst effects are frequently a by-product of acid mine drainage.[17]

Manganese. Manganese compounds are more readily solubilized than those of iron; *i.e.*, a less reducing environment is required. Objections to discharges of this metal (*e.g.*, staining properties) are similar to those for iron. Not a toxicologically significant metal.

Nickel. Corrosion from stainless steels, etc. Toxic to fish.

Titanium. Objectionable in suspended solid form. Not toxic to fish.

Zinc. Ubiquitous associate of industrialization. Impact closely parallels that of copper.

In those industrialized regions of the world where aqueous heavy-metal pollution--real or imaginary-- has become a major official concern, legislative controls have been applied, somewhat haphazardly, to both water resources and to the associated biota. The most stringent requirements are, not unreasonably, concerned with drinking water quality and standards as illustrated in Table 2.4. Implementation of these programs has generally suffered in the past from a dearth of data (see Reference 18). With the current accelerating use[19] of the various modern trace analysis techniques (as discussed in this volume and elsewhere) by the governmental monitoring agencies, there should soon be no lack of information regarding sources and areas of "pollution"; maintaining quality within the legal limitations may be a little more troublesome, however.

Table 2.4

Drinking Water Quality (μg/ℓ)

	a	*b* Max.	*b* Mean	*c*	*d*
Ag	n.d.	7	0.2	50[f]	
As				50[f]	200
Bi	n.d.				
Cd	n.d.			10[f]	50
Co	n.d.	9.5			
Cr	5	35	0.4	50[f]	50
Cu	20	250	8	1,000	1,000
Fe	100[e]	1,700	43	300	300
Ga	n.d.				
Hg				5[f]	
Mn	10[e]	1,100	1	50	100
Mo	100	68	68		
Ni	n.d.	34	3		
Pb	n.d.	62	4	50[f]	100
Sn	n.d.				
Ti	2				
V	n.d.	70	4		
Zn	10			5,000	5,000

[a] Reported concentrations for one specific public water supply. (Data from Reference 18.)

[b] Concentrations in finished water of 100 largest U.S. cities in 1962. (Data from Reference 20.)

[c] 1962 U.S. Public Health Service standards.

[d] 1958-61 World Health Organization standards

[e] Data for unfiltered samples

[f] Mandatory maximum concentrations.

The situation with regard to the heavy-metal contents of the aqueous food species is considerably more controversial. Tables 2.5 and 10.6 give some typical "fish toxicity" data culled from the literature. The definitions of "lethality" and "toxicity" used for such tabulations are open to considerable criticism (see Chapter 9). Although

Table 2.5

Aquatic Life Tolerance to Heavy Metals in Freshwater
(Data from Reference 21)

Metal	Threshold Concentration[a] (µg/ℓ)	Lethal Concentration[b] (µg/ℓ)
Ag	10	
As	1,000	
Cd	10	300
Cr	50	
Cu	20	20
Hg	10	10
Ni	50	1,000
Pb	100	300
Zn	100	300

[a]Concentration value safe for most fish species if not exceeded.

[b]Concentration lethal to one particular common fish; nonstandardized exposure times.

this type of data may be for total or unspecified portions of the organism, metal uptake is selective for the constituent components of the animal (see Table 10.6) and man in turn consumes only specific portions of many of these species. Considerations such as these, coupled with the fact that frequently fish and shellfish do not constitute a major portion of man's diet, argue for considerable caution in imposing strict legal limits on the allowable metal contents of these foods. In any event, the analyst carries a heavy responsibility.

In many parts of even highly developed nations, treatment of public water supplies is still rudimentary, so that the widespread introduction of innovative tertiary treatment processes specifically designed to remove or recover heavy-metal species probably lies far in the future. Several viable pilot-plant processes have been developed, however. Of these, reverse osmosis[22] and nonspecific sorption mechanisms (for example, onto a sulfide substrate[23]) offer the most promise. It has been demonstrated[24] also that charcoal filtration will remove, for example, some 80% of the soluble arsenic content from a water supply.

Characterizing Dynamic Systems

The problems inherent in efforts to adequately describe or predict the behavior of nonstatic natural water systems are touched on briefly in Chapter 4. Very clearly, the formulation of a sampling program addressed to well thought out and specific problems is of paramount importance. Considering only the impingement of "point-source" metal pollution on man, two basic sampling programs may be recommended:

1. Sampling at some distance from source, within the receiving rivers, lakes or oceanic areas. In this fashion, the resultant data should indicate (in general terms) potential exposure within that ecosystem.
2. Direct measurement of the effluent at source. This is the current focus, for example, of the U.S. Environmental Protection Agency monitoring program.[25]

ANALYTICAL REQUIREMENTS

Clearly the analytical technique selected for the determination of the aqueous environment heavy metals must be capable not only of detecting extremely low concentrations but of quantifying perturbations within well-understood statistical limits. Conversely we should aim to better our knowledge of the natural and stressed distribution of these metals by continually developing capabilities for resolving ever-diminishing quantities and differences. Within the context of the current social concern with environmental quality the gaps in our knowledge are frightening, but it is well worth the trouble to pause occasionally to consider the implications of analysis in the sub-ppm range. Thus whereas for "high level" concentrations an analytical method may be selected based upon

criteria such as relative rapidity or suitability
for direct field analysis, the scope of a trace-metal
monitoring program is usually directly circumscribed
by the capabilities of the available instrumental
techniques.

Natural Water Analysis

First and foremost, wherever possible, methodology
directed at the trace constituents in water should
be applicable to solutions *per se*, since it is of
paramount importance to strictly curtail all pre-
analysis sample treatments. The following considerations (discussed elsewhere) will, to a lesser or
greater degree, depending upon individual circumstances,
shape the structure of the final analytical program:

1. problems associated with the *dynamics* of the
   water system
2. potential matrix interferences
3. effects of (chiefly) organic complexation
4. coexistence of various inorganic and organic
   solid phases.

Analysis of Nonaqueous Phases

It is assumed that the environmental analyst
will, at some stage, be concerned with the composition
of naturally occurring nonaqueous phases, although a
detailed consideration of this topic is not feasible
within the scope of this book. The trace-metal
contents of the marine and freshwater biota are of
particular importance and some individual examples
are cited in Chapter 10. Analysis by AS methods
necessitates prior dissolution (Chapter 5), but
absolute concentrations are generally considerably
enhanced over the coexisting aqueous values,
alleviating somewhat many of the procedural difficulties. It may be possible in some instances to
gain an insight into the composition of the water
from a knowledge of the metal contents of selected
organisms. It is unlikely, however, that any very
precise data will emerge from such a postulated use
of the natural concentrating abilities of the
organisms, given our general ignorance of the
mechanisms involved and the near impossibility of
defining the water observed over any particular
time period.

The impingement of hydrocarbon and other organic residues on natural water bodies will undoubtedly be a major concern in many laboratories. Considerable work has been accomplished with regard to the "finger printing" of various petroleum products using trace-metal spectra (in association with other parameters), but little is known of possible modifications to natural trace-metal budgets posed by spills and leakages of these products into the aqueous environment. Solubilization of metals such as nickel and vanadium has been postulated as a likely outcome,[26] but no definitive data are as yet available. Organic constituents may be separated directly or *via* suitable solvents and nebulized directly into flame AS equipment. Westwood[27] has discussed the use of emulsions and dispersions to facilitate this operation. Precise standardization has suffered in the past from the lack of suitable organometallic standards, but this problem has been ameliorated considerably in recent years.

Application of Solid-Sample Analysis Techniques

It is theoretically possible to use very many chemical analysis techniques to measure the heavy-metal constituents of water, provided that they are preceded by the necessary individual processing treatments. But it should be a prime aim of the water analyst to drastically reduce reliance on the latter in order to improve the accuracy and precision of the final data. Unfortunately it is a natural tendency to apply methods better suited either to solid samples or to samples containing considerably higher concentrations of the analyte metal where equipment for such methods is readily available to the analyst. Prime examples of this undesirable practice--in this case the use of conventional X-ray fluorescence spectroscopy and arc spectrography--are given in Table 2.7 (see also Reference 29). The concentrations required in these cases should be noted in particular; concentration factors up to X5000[30] are not unknown. These and similar techniques offer some real advantages for biota and sediment analysis, but routine application to natural aqueous samples should be discouraged.

Table 2.6

Two Examples of "Solid-Sample" Analytical Techniques
Applied to Heavy Metals in Natural Waters

Technique	spectrographic	X-ray fluorescence
Sample type	freshwater	freshwater
Reference	5, 27	28
Concentration Range in Raw Sample	$\sim$ 1 µg/ℓ	$\sim$ 5 µg/ℓ
Initial Sample Size	3500 mℓ	1000 mℓ
Concentration Procedure	cocrystallization/ precipitation	chelations/solvent extraction
Total Concentration Factor	X 3500	X 2-4000

Standard Methods and Colorimetry

"Standard methods" are immensely important in
metal pollution control since these are the designated
techniques legally accepted in official disputes.
Because of the need for widespread applicability and
interlaboratory standardization, only certain, well-
tried and relatively simple procedures are adopted;
thus a "cookbook" approach to each analysis may be
instituted, negating the necessity for highly skilled
technical help. To this end, various long-established
colorimetric procedures have most usually been
sanctioned, and new procedures are only very cautiously
introduced after exhaustive evaluations. Table 2.7
tabulates the active reagents required for certain
colorimetric procedures (including some designated
"standard methods") in common use for the analysis
of freshwater samples. Many of these reactions are
nonspecific and require various separations or the
addition of masking reagents. The interferents
included in Table 2.7 should be carefully noted,
but reference must be made to one of the standard
texts[31-33] for specific procedures. The final
quoted detection limits attainable are frequently
impressive (see Table 2.13), but it must be remembered

Table 2.7

Some Standard Colorimetric Reagents for Heavy-Metal
Pollutant Analysis of Fresh Natural and Waste Waters

Element	Reagent[a]	Major Interferents[b]	pH[c]
Ag[d]	dithizone[e]	Cu, Hg	acid
As	Ag–DEDTC[e]	Sb	acid
Cd	dithizone[fg]	Au, Co, Cu, Hg, Ni	alkali
Cr (VI)	diphenylcarbazide[efg]	Hg, Mo	acid
Cu	dithizone	Ag, Hg	acid
Cu	DEDTC[f]	Bi, Co, Fe, Ni	alkali
Cu (I)	cuproine	Fe, Ni	neutral
Cu (I)	bathocuproine[eg]		neutral
Cu (I)	neocuproine[hg]		neutral
Cu (I)	cuprethol[e]	various	acid
Hg	diphenylcarbazine	Mo, Cr	acid
Hg	dithizone[f]	Ag, Cu	
Fe	thioglycollic acid[f]	Co, Ni, V	alkali
Fe (II)	dipyridyl	various	acid
Fe (II)	phenanthroline[h]		acid
Fe (II)	tripyridine[e]		alkali
Fe (III)	sulphosalicylic acid	Ti	acid
Fe (III)	thiocyanate	various	acid
Mn[d]	(permanganate)[efg]		acid
Mn	formaldoxime	Cu, Fe	alkali
Ni	heptoxime[gh]	Cu, Fe	acid
Ni	dimethylglyoxime[ef]	Fe	alkali
Ni	nioxime	Fe	alkali
Ni	DEDTC	Bi, Co, Cu, Fe	alkali
Pb	dithizone[fhg]	Bi, Sn, Th	alkali
Se	diaminobenzidine[e]		alkali
Th	thorin[h]		
Ti (IV)	tiron	Fe, Mo, U	neutral
V	oxine	Cu, Fe, Ti	acid
Zn[d]	dithizone[efg]	Ag, Cd, Cu, Hg, Pb	acid
Zn	zincon[e]	Cr, Cu, Fe, Mn, Ni	alkali

[a] Nonchemical name utilized where in common use (see Appendix 2).
[b] Major common interferents shown only. See standard texts[31,32] for specific reactions.
[c] pH range indicated. This parameter is critical for each reaction and for masking interferents.
[d] Atomic absorption as tentative standard method under certain conditions.
[e] ASTM standard method in U.S.A. (Reference 34).
[f] ABCM–SAC joint committee standard method in U.K.
[g] ASTM recommended method for heavily polluted waters.
[h] Tentative ASTM standard method.

that these values are with reference to the final
sample solution after extraction/separation treatment.
AA analysis is gradually replacing the standard
photometric procedures for cobalt, iron and zinc,
and tentative procedures have been developed for
several other metals including arsenic and selenium.

Within the limitations noted, colorimetry is a
useful tool for routine water analysis; new techniques
are being constantly introduced. These latter usually
involve development of more specific rather than more
sensitive reagents. Table 2.8 cites recent literature
applications to trace metals in seawater.

Table 2.8

*Some Colorimetric Reagents Suitable for
Heavy Metals in Seawater*

Element	Reagent[a]	Reference
Co	nitroso - R	35
Cu	neocuproine	36
	PHTTT	37
Fe	bathophenanthroline	38
Mn	formaldoxime	39
Mo	dithiol	40
Ni	dimethylglyoxime	41
	OXTD	42
Sn	phenylfluorone	43
W	dithiol	40
Zn	dithizone	31

[a]See Appendix 2.

Methods Especially Suited
to Solution Analysis

The analytical procedures best suited to natural
water analysis are those which utilize solutions
directly. This is clearly so where direct analysis
is possible, as illustrated by Table 2.9. Even where
pre-analysis concentrations are mandatory, the most
common of such procedures (solvent extraction, resin
exchange; see Chapter 5) yield solutions as end
products. The following methods are advantageous.

Table 2.9

Characteristics of a Trace-Metal Technique Suited to
Natural Water Analysis*   (Data from Reference 44)

Elements Determined	Cd, Cu, Pb, Zn
Technique	anodic stripping voltammetry
Sample Type	natural water
Concentration Range in Raw Sample	$10^{-7} - 10^{-8}$ M (depending on electrode system)
Initial Sample Size	15-mℓ
Concentration Procedure	not required

*Compare with Table 2.6.

Colorimetry

Colorimetry (photometry) has been considered in the previous section.

Ion-Selective Electrodes

These membrane electrodes operate in a fashion analogous to pH electrodes[45],[46] and are currently available for a restricted number of heavy metals (Table 2.10). They have generated considerable interest for a range of water analysis applications[47-49] since the direct, rapid and continuous readout capabilities offer the required characteristics for

Table 2.10

Ion-Selective Electrodes Available for Heavy-Metal Analysis
(Data from References 45, 51, 52)

Membrane Type	Element	Detection Limit (M)	Major Interferents
Solid	Ag	10^{-8}	Hg
	Cd	10^{-7}	Ag, Cu, Hg
	Cu	10^{-8}	Ag, Hg, S
	Pb	10^{-7}	Ag, Cu, Hg

in situ monitoring programs. However, there are a
number of interferences which must be allowed for,
and *activities* rather than concentrations are
determined. Thus not only must the metal be present
in an "available" form, but only dilute or constant
ionic strength solutions should ideally be measured.

Polarography

Polarographic techniques have been widely applied
to natural waters (see review by Maienthal and Taylor[52]
and see also Table 2.11) and anodic stripping voltammetry in particular has proved to be advantageous
for the direct analysis of these waters since the
incorporated concentration step enhances the detection limit capability. In addition, some of the
metals of most concern as potentially troublesome
pollutants (cadmium, copper, lead and zinc) are
especially amenable to determination by this method
(Table 10.5). Anodic stripping analyzes "free"
ionic species of the metal in solution, but the
complexed fraction may be determined following acid
treatment as illustrated in Table 2.2.

Table 2.11

Polarographic Techniques Available for Trace Heavy-Metal Analysis
(Data from References 44, 52-55)

Technique	Electrode	Detection Range (M)
Conventional Polarography	M.D.E.	10^{-5}
Pulse Polarography	M.D.E.	$10^{-7} - 10^{-8}$
Cathode-Ray Polarography	M.D.E.	10^{-7}
Anodic Stripping Voltammetry	M.D.E.	$10^{-8} - 10^{-10}$
Anodic Stripping Voltammetry	G.E.	$10^{-7} - 10^{-10}$

M.D.E.--mercury drop electrode
G.E.   --graphite electrode

Neutron Activation Analysis

This type of analysis may be applied to both liquid and solid samples and is a very sensitive technique for a very wide range of metals (see Tables 2.12 and 2.13), although the required equipment is exceedingly expensive and complex and is seldom readily available to the average investigator.

Table 2.12

Ultra-Trace Metal Analysis Techniques: A Comparison of Absolute Detection Limits (pg) for the Major Pollutant Metals[a]

	Atomic Absorption[b]	Atomic Fluorescence[b]	Polarographic[c]	Neutron Activation[d]
Ag	2×10^{-1}	1		5×10^3 [k]
As	10^2	5×10^2	10^4 [h]	10^3
Cd	10^{-1}	2×10^{-1}	2×10^4	5×10^4
Cu	7	5	1.5×10^3	10^3
Hg	10^2 [e]	6×10^2 [f]	2×10^4 [i]	4×10^3 [l]
Pb	5	10	5×10^3	5×10^5
Sb	30	2×10^2 [g]	1.5×10^7 [j]	2×10^3
Zn	8×10^{-2}	2×10^{-2}	8×10^3	3×10^4

[a]See text for limitations on intertechnique comparisons.
[b]Carbon filament data from Table 7.8 except as noted.
[c]Anodic stripping voltammetry data from References 44 and 55 except as noted; 25-ml sample.
[d]Data from Reference 56 except as noted; 8-hr irradiation; 10^{13} flux.
[e]Cold vapor absorption, Hatch and Ott[57] method; LDC Inc. equip.
[f]Reference 58.
[g]Reference 59.
[h]Data from Reference 60; single sweep polarography.
[i]Reference 61; 25-ml sample.
[j]Data from Reference 62; cyclic voltammetry; 25-ml sample.
[k]Reference 63; 1-hr irradiation; 10^{12} flux.
[l]Reference 64; requires radiochemical separation.

Table 2.13

Detection Limits (µg/ℓ) by Trace Analysis Techniques
Other than AA, AF or AE[a]

Element	Spectrographic arc[b]	spark[c]	Colorimetric[d]	Neutron Activation	Fluorescence
Ag	10	10	5	0.5	
As	g	g	10	2	8
Au	50		3	0.05	
Bi	50	g	10	50	
Cd		200	3	5	
Co	10	500	2	0.5	
Cr	20	300	7	100	
Cu	20	50	0.8	0.1	2
Fe	100	200	2	g	1
Ga	20		40	0.5	5
Hg	70	g	5	1	
In	200	g	50	0.01	
Mn	20	20	5	0.005	
Mo	20	300	10	10	
Ni	20	500	4	5	2
Is			50	5	
Pb	20	g	6	g	
Pd			10	5	
Pt		g	3	5	
Re			200	0.1	
Rh	g	700	50	0.05	
Ru		g	100	1	
Sb	200	g	4	0.3	
Sn	70		100	50	
Ti		100	10	5	0.6
Tℓ	500	g	20		
V		200	10	0.1	0.7
W			30	0.1	
Zn		2			

[a]Data for discrete sampling mode (both liquid and solid samples)
standardized to solution analysis units. Values poorer than
1000 µg/ℓ omitted.

[b]Data from References 65, 66; minimal line blackening--see these
sources for other definitions and limitations.

[c]Data from References 67 (rotating disc; detection limit defined
as x 1.3 adjacent background) and 68 (porous cup; detection
limit not defined).

[d]Data from References 69, 70 (1-mℓ solution in 1-cm cell). The
values cited are for concentrations in the final solution after
extraction.

[e]Data based on References 56, 63, 71 recalculated to approximate
10-mℓ sample; 1-hr irradiation, 2×10^{12} flux.

[f]Proton-induced X-ray fluorescence data from Reference 72; 1-mg
sample.

[g]Values lie between $10^3 - 10^4$ µg/ℓ.

One major difficulty in applying this mode of analysis to natural aqueous samples (seawater samples in particular) is the formation and subsequent masking action of ^{24}Na, so that complex pre- and post-irradiation separations are usually mandatory.

Atomic Spectroscopy

This method of analysis--the subject of this book--is covered in some detail in the following chapter.

Ultra-Trace Analysis

Only a very few analytical methods are considered to be well suited to the determination of metals present in ultra-trace (*i.e.*, ng-pg range) amounts in water. Such procedures should provide extremely superior sensitivity and detection limit character- istics so that pre-analysis manipulations of the sample are not limiting. Table 2.12 lists techniques offering the best combination of advantages for water analysis based on these criteria and gives, for each technique, the absolute detection limit values for some major pollutant metals. It should be noted that since these data are in weight units, direct com- parison between the various procedures is not possible. For example, the filament-atomization mode AS analysis techniques (see Chapter 7) included in this tabulation utilize µℓ range samples so that the resultant *concentration* values are not necessarily outstanding.

Automated Analysis

Many analytical methods may be modified to speed the processing of *routine* samples, although effort is often expended on inherently poor procedures.[73] Simple manipulations in this category include use of coupled automatic sample changers (see following chapter) and on- or off-line electronic data pro- cessing. However, for freshwater pollution monitoring programs in particular (*i.e.*, within potentially rapidly fluctuating environments), there is an urgent need for *in situ* continuous monitoring modules.[74,75] Such systems for dissolved oxygen contents, conduc- tivity, pH and turbidity parameters are now being widely deployed but the necessary technology for trace-metal detection has not yet been well developed.

Specific ion electrodes appear to offer the most promise for the future as *in situ* monitors,[45] although the interference effects from concomitants noted above cause considerable practical complications. Portable and rugged electrodes are available for silver, cadmium, copper and lead, and suitable probes for other elements are certain to become available with time. To date, more work on automated chemical analysis has been accomplished using colorimetric procedures, and particularly through use of the Technicon Auto-Analyzer system. Only those heavy metals likely to be present in "natural" waters in relatively high concentrations (*e.g.*, iron and manganese) have been studied in any detail,[6] but many metals commonly concentrated in industrial waste discharges (*e.g.*, cadmium, chromium, copper, nickel, zinc) are potentially determinable (*e.g.*, Reference 76). An extension of this proprietary system to AF analysis is cited in the following chapter.

Summary of Techniques

Table 2.13 gives published detection limits for non-AS trace analytical techniques applicable to a wide spectrum of heavy metals. This table is included by way of comparison with the AS data of Tables 3.5 and 3.6, but direct, uncritical comparisons should be applied with the greatest caution, and the reader must be aware of different usages of the terms "detection limit" and "sensitivity." The techniques of Table 2.13 exhibit wide universality, but this may not be a major concern of the water analyst. Optimum ultra-trace methods for the metals likely to be of prime interest in most water pollution laboratories have been given in Table 2.12. The characteristics of AS techniques are discussed at length in Chapter 3. The following briefly summarizes the relevant attributes of the remaining recommended trace analysis procedures considered above.

Universality and Selectivity

The techniques of Table 2.13 have wide utility. The X-ray fluorescence data listed are not of the conventional type since heavy particles were used as an excitation source.[72,77] This is a relatively new development and many other elements, for which detection limit data are not yet available, are potentially determinable in ultra-trace quantities.

Table 2.14

Interlaboratory Study of Precision and Accuracy of Several
Analytical Techniques for Aqueous Heavy Metals (μg/ml)[a]

Element	Conc.	Method	Number[c]	Rel. Stand. Deviation[d]	Rel. Error[e]
Ag	0.15	dithizone[f]	14	61.0	66.6
		spectrography	3	73.5	40.0
As	0.04	Ag-DEDTC[f]	46	13.8	0.0
Cd	0.05	dithizone[f]	44	24.5	0.0
		polarography	4	68.0	20.0
Cr	0.11	permanganate[f]	31	47.8	27.2
		spectrography	4	37.6	36.3
		polarography	3	18.4	27.2
Cu	0.47	DEDTC[f]	12	11.7	6.0
		spectrography	3	14.0	14.9
		atomic absorption	3	2.3	2.1
		polarography	2	22.5	19.1
Fe	0.30	phenanthroline[f]	44	25.5	13.3
		spectrography	5	32.2	6.6
		atomic absorption	2	26.4	0.0
Mn	0.12	persulfate[f]	33	26.3	0.0
		spectrography	4	19.4	8.3
		atomic absorption	3	19.6	8.3
Pb	0.07	dithizone[f]	43	42.1	8.5
		spectrography	3	62.4	32.8
		atomic absorption	2	8.2	21.4
Se	0.02	diaminobenzidine[f]	35	21.2	5.0
V	0.006	gallic acid	22	20.0	0.0

[a]ARS Water Metals No. 3, study 23: Ag, Cd, Cr, Cu, Fe, Mn, Pb;
Water Trace Elements No. 2, study 26: As, Se, V   (Data from
Reference 78).
[b]Given concentration of synthetic samples.
[c]Number of independent determinations utilized.
[d]Standard deviation as % of the mean representing the data
precision.
[e]Mean as % of true values representing the accuracy.
[f]Standard method in U.S. (Reference 34).

The polarographic and specific-ion electrode techniques have the most restricted use. Selectivity basically refers to analytical interferences from concomitants. Unfortunately, it is not possible to adequately summarize this topic here; the interested reader is urged to consult the standard cited literature.

Speed, Ease and Cost

Disregarding other factors, use of the specific-ion electrodes is the least, and neutron activation the most, expensive mode of analysis. However, the true cost of any particular program is not to be measured in terms of the cost of the basic equipment alone. Speed and flexibility are important considerations with regard to continuous or automated systems.

Precision and Accuracy

This important topic dominates the contents of Chapter 9. Very few good intertechnique comparisons are available but one such comparison--a study basically concerned with existing or potential "standard methods"--is summarized in Table 2.14.

REFERENCES

1. National Academy of Sciences. *Marine Environmental Quality,* Ocean Sci. Comm., Nat. Acad. Sci., Washington, D.C., 1971.
2. Goldberg, E. D. in *Chemical Oceanography* (J. P. Riley and G. Skirrow, eds.), Academic Press, London, 1965, pp. 163-196.
3. Goldberg, E. D. in *The Sea,* Vol. 2 (M. N. Hill, ed.), Interscience Publishers, New York, 1963, pp. 3-25.
4. Turekian, K. K. and M. D. Kleinkopf. *Bull. Geol. Soc. Am., 67,* 1129 (1956).
5. Silvey, W. D. *Geol. Surv.,* Water Supply Paper 1535-L, U.S. Gov't. Printing Office, Washington, D.C., 1967.
6. Mancy, K. H. (Editor). *Instrumental Analysis for Water Pollution Control,* Ann Arbor Science Publishers, Ann Arbor, Mich., 1971.
7. Rashid, M. A. *Soil Sci., 3,* 298 (1971).
8. Rashid, M. A. and J. D. Leonard. *Chem. Geol., 11,* 89 (1973).

9. Murray, C. N. and L. Murray. IAEA Symposium on the Interaction of Radioactive Contaminants with the Constituents of the Marine Environment, Seattle, Wash., July, 1972.
10. Piro, A., M. Bernhard, M. Branica, and M. Verzi. IAEA Symposium on the Interaction of Radioactive Contaminants with the Constituents of the Marine Environment, Seattle, Wash., July, 1972.
11. Windom, H. L. Meeting Preprint 1572, American Society of Civil Engineers, 1972.
12. Williams, R. and M. Murdock. in *Proceedings of the Second National Symposium and Radioecology* (D. J. Nelson and F. C. Evans, eds.), Springfield, Va., 1969, pp. 431-439.
13. David, E. L. *Water Res. R.*, *7*, 453 (1971).
14. Stumm, W. and E. Stumm-Zollinger. in *Nonequilibrium Systems in Natural Water Chemistry* (J. D. Hem, ed.), American Chemical Society, Washington, D.C., 1971.
15. Yamagata, N. and I. Shigematsu. *Bull. Tokyo Inst. Pub. Health*, *19*, 2 (1970).
16. Irukayama, K. *Advan. Water Poll. Res.*, *3*, 153 (1966).
17. Barnes, H. L. and S. B. Romberger. *J. Water Pollution Control*, *40*, 371 (1968).
18. Barnett, P. R., M. W. Skongstad, and K. J. Miller. *J. Am. Water Works Assoc.*, *61*, 61 (1969).
19. Stack, V. T. *Anal. Chem.*, *44*, 32A (1972).
20. Durfor, C. N. and E. Becker. *Geol. Surv.*, Water Supply Paper No. 1812, U.S. Gov't. Printing Office, Washington, D.C., 1964.
21. Todd, D. K. (Editor). *The Water Encyclopedia*, Water Information Center, Port Washington, N.Y., 1970.
22. Linstedt, K. D., C. P. Houck, and J. T. O'Connor. *J. Water Pollution Control.*, *43*, 1507 (1971).
23. Kraus, K. A. and H. O. Phillips. U.S. Patent 3,317,312 (1967).
24. Angino, E. E., L. M. Magnuson, T. C. Waugh, O. K. Gallee, and J. Bredfeldt. *Science*, 168 (1970).
25. Ballinger, D. G. *Environ. Sci. Technol.*, *6*, 130 (1972).
26. Bugel'skii, Y. Y. and L. S. Tsimliyanskaya. *Kora Vyvetrivaniya, Akad. Nauk SSSR*, *7*, 148 (1966).
27. Silvey, W. D. and R. Brennan. *Anal. Chem.*, *34*, 784 (1962).
28. Marcie, F. J. *Environ. Sci. Technol.*, *1*, 164 (1967).
29. Tackett, S. L. and M. A. Brocione. *Anal. Lett.*, *2*, 649 (1968).
30. Morris, A. W. *Anal. Chim. Acta*, *42*, 397 (1968).
31. Sandell, E. B. *Colorimetric Determination of Traces of Metals*, Interscience Publishers, New York, 1958.
32. Snell, F. D. and C. T. Snell. *Colorimetric Methods of Analysis*, Vol. IIA, D. Van Nostrand, New York, 1959.

33. Pinta, M. *Detection and Determination of Trace Elements,* (Translated from the French by M. Bivas), Ann Arbor Science Publishers, Ann Arbor, Mich., 1971.
34. American Public Health Association. *Standard Methods for the Examination of Waste and Wastewater,* 12th ed., New York, 1965.
35. Shipman, W. H. and J. R. Lai. *Anal. Chem., 28,* 1151 (1956).
36. Jones, P. D. and E. J. Newman. *Analyst, 87,* 637 (1962).
37. Stiff, M. J. *Analyst, 97,* 146 (1972).
38. Riley, J. P. and H. Williams. *Mikrochim. Acta, 35,* 804 (1959).
39. Bradfield, E. G. *Analyst, 8,* 254 (1957).
40. Kawabuchi, K. and R. Kuroda. *Anal. Chim. Acta, 46,* 23 (1969).
41. Forster, W. and H. Zeitlin. *Anal. Chim. Acta, 35,* 42 (1966).
42. Forster, W. and H. Zeitlin. *Anal. Chem., 38,* 649 (1966).
43. Smith, J. D. *Analyst, 95,* 347 (1970).
44. Allen, H. E., W. R. Matson, and K. H. Mancy. *J. Water Pollution Control., 42,* 573 (1970).
45. Durst, R. A. National Bureau of Standards Publication 314, U.S. Gov't. Printing Office, Washington, D.C., 1969.
46. Durst, R. A. *Am. Scientist, 59,* 353 (1971).
47. Andelman, J. B. *J. Water Pollution Control, 40,* 1844 (1968).
48. Whitfield, M. AMSA Handbook No. 2, Australian Marine Science Assoc., Sydney, 1971.
49. Coleman, D. M., R. E. VanAtta, and L. N. Klatt. *Environ. Sci. Technol., 6,* 452 (1972).
50. Brand, M. J. D., J. J. Militello, and G. A. Rechnitz. *Anal. Lett., 2,* 523 (1969).
51. Rechnitz, G. A. and N. C. Kenny. *Anal. Lett., 3,* 259 (1970).
52. Maienthal, E. J. and J. K. Taylor. in *Trace Inorganics in Water,* (R. A. Baker, ed.), Advan. Chem. Ser. No. 73, American Chemical Society, Washington, D.C., 1968, pp. 172-182.
53. Kemula, W. *I.U.P.A.C. Int. Symp. on Analytical Chemistry* (A. M. G. MacDonald and W. I. Stephen, eds.), Butterworths, London, 1970.
54. Whitnack, G. C. and R. Sasselli. *Anal. Chim. Acta, 47,* 267 (1969).
55. Zirino, A. and M. L. Healey. *Environ. Sci. Technol., 6,* 243 (1972).
56. Guinn, V. P. unpublished data (1971).
57. Hatch, W. R. and W. L. Ott. *Anal. Chem., 40,* 2085 (1968).
58. Muscat, V. I., T. J. Vickers, and A. Andren. *Anal. Chem., 44,* 218 (1972).
59. Massman, H. *Spectrochim. Acta, 23B,* 215 (1968).

60. Whitnack, G. C. and R. G. Brophy. *Anal. Chim. Acta, 48,* 123 (1969).

61. Perone, S. P. and W. J. Kretlow. *Anal. Chem., 37,* 968 (1965).

62. Jacobsen, E. and T. Rojahn. *Anal. Chim. Acta, 54,* 261 (1971).

63. Guinn, V. P. and H. R. Lukens. in *Trace Analysis: Physical Methods* (G. H. Morrison, ed.), John Wiley and Sons, New York, 1965, pp. 325-376.

64. Weiss, H. V. and T. E. Crozier. *Anal. Chim. Acta, 58,* 231 (1972).

65. Addink, N. W. H. in *U.S. National Bureau of Standards Monograph 100* (W. W. Meinke and B. F. Scribner, eds.), Washington, D.C., 1967, pp. 121-148.

66. Kaiser, H. in *U.S. National Bureau of Standards Monograph 100* (W. W. Meinke and B. F. Scribner, eds.), U.S. Gov't. Printing Office, Washington, D.C., 1967, pp. 149-164.

67. Baer, W. K. and E. S. Hodge. *Appl. Spectrosc., 20,* 281 (1966).

68. Feldman, C. in *Conference on Analytical Chemistry and Applied Spectroscopy,* Pittsburgh, 1963.

69. Hume, D. N. in *Equilibrium Concepts in Natural Water Systems* (R. F. Gould, ed.), American Chemical Society, Washington, D.C., 1967, pp. 30-44.

70. Morrison, G. H. and R. K. Skogerboe. in *Trace Analysis: Physical Methods* (G. H. Morrison, ed.), Interscience Publishers, New York, 1965, pp. 1-24.

71. Winchester, T. W. *Progr. Inorg. Chem., 2,* 1 (1960).

72. Johnson, T. B., R. Akelsson, and S. A. E. Johansson. Unpublished manuscript.

73. Steinder, R. L., J. B. Austin, and D. W. Lander. *Environ. Sci. Technol., 3,* 1192 (1969).

74. Blaedel, W. J. and R. H. Laessing. *Advan. Anal. Chem. Instru., 5,* 69 (1966).

75. Gafford, R. D. in *Proceed. FAO Tech. Conf. on Marine Poll.,* Rome, 1971.

76. Berry, A. J. in *Automation in Analytical Chemistry: Technicon Symposium 1966,* Vol. 1, Mediad, New York, 1967, pp. 560-571.

77. Johnson, T. B., R. Akelsson, and S. A. E. Johansson. *Nucl. Instr. Methods, 84,* 141 (1970).

78. McFarren, E. F., and R. J. Lishka. in *Trace Inorganics in Water* (R. A. Baker, ed.), Advan. Chem. Ser. No. 73, American Chemical Society, Washington, D.C., 1968, pp. 253-264.

CHAPTER 3

ATOMIC SPECTROMETRY

"Flame methods, properly applied, now provide results for a number of elements by which all other
methods must be judged, especially at low levels."

E. E. Pickett and S. R. Koirtyohann[1]

BASIC PRINCIPLES

The basic reaction underlying atomic spectrometry
may be simply stated:

$$R^\circ + h\nu \rightleftarrows R^{\circ *} \qquad (3.1)$$

The ground-state atom (R°; see symbol key in Appendix
1) absorbs energy to yield the excited state and
emits radiation following de-excitation. For atomic
fluorescence, both the forward and reverse reactions
of Equation 3.1 apply.

Atomic Emission

The radiation emitted by the reverse reaction of
Equation 3.1 depends upon the specific transition
involved (for example, the transition from excited
level j to the ground state). The thermal emission
intensity for this particular transition would then
be:

$$I_E = \frac{N_j E_j}{\tau} \qquad (3.2)$$

47

Note that the radiation intensity depends upon N_j--
the number of atoms populating the excited j level--
as described by the Boltzmann equation:

$$N_j = N_o \cdot \frac{g_j}{Z} \cdot e^{-E_j/kT} \qquad (3.3)$$

where the atomic partition function (Z) is given by
$\Sigma g \exp - E/kT$ for all possible transitions (g is a
statistical weight term; see Reference 2). Since
$E_j = h\nu_j = hc/\lambda_j$, it follows that for a fixed T,
N_j/N_o increases with λ.

This latter ratio [the Boltzmann distribution
is fundamental to all atomic spectrometric (AS)
procedures] is very small over the temperature ranges
of typical flames (3-4,000°C) and has a value of
around 10^{-4} - 10^{-10}. For sodium at 3,000°K, for
example, N_j/N_o has the value 6 x 10^{-4}. Figure 3.1
shows the zinc distribution over operable temperatures,
compared with a more readily excited alkali earth
element. Since atomic absorption (AA) spectrometry

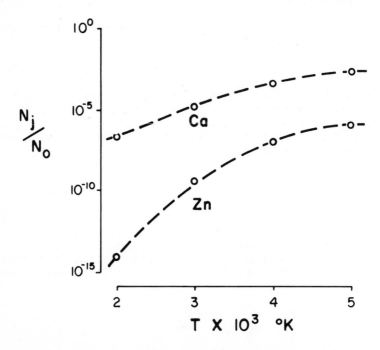

Figure 3.1. *Atoms available for absorption and fluorescence:*
            N_j/N_o as a function of temperature for two
            elements. (Data from References 3 and 4.)

is dependent upon the ground-state atom population
(N_O), whereas atomic emission (AE) results from de-
excitation processes (and is thus a function of N_j),
the general order of magnitude of the N_j/N_O ratio
is commonly cited as proof of the fundamental
superiority of AA over AE. The optical parameters
measured in both methods are quite different, and
Alkemade[5] has discussed in detail the fallacies
inherent in such simplistic reasoning. The potential
superiority of AA depends upon the source intensity
(radiance) as compared with a theoretical black-body
radiator at the flame temperature (T) and line wavelength (λ) used; it has been noted[1] that the comparison depends, in fact, upon T and λ only, and not
upon the N_j/N_O ratio. AA is theoretically the
superior technique only where long wavelength
resonance lines are used. It should be noted, however, that minor temperature fluctuations have a far
greater effect upon N_j than N_O (Equation 3.3), so
that AE is quite sensitive to this parameter; thus
appropriate safeguards must be taken.

Line Profiles

    The type of transitions--to and from the atomic
ground state--considered so far give resonance
analysis lines. These lines, because they involve
ground-state atoms, are the most important for AS
analysis, although the atom may be excited to higher
and higher states until, at the ionization energy
threshold, the electron is removed to form an ion.
Ionization energies are included later in Table 3.3.
Some of the higher energy transitions, and even ion
lines, are useful in AE analysis.
    The emission lines are not perfectly monochromatic.
The basic natural line width--around 10^{-4} Å--is due
to the finite lifetime (τ) in the excited state.
Various additional physical processes act to further
broaden the line, and the resulting line width is
generally defined as the value (λ) at one-half peak
height. In AE, the emission line width is not
generally thought to be of much importance since it
is usually considerably narrower than the conventional
monochromator band-pass (but see Reference 6 for a
consideration of echelle gratings). AA and AF
analysis, however, depend upon the subtraction of
energy from a source emission. Since the monochromator
resolution is usually poorer than either the emission
or the absorption line width, a practical (line-source)
analysis system has to depend upon the absorption

line width being much broader than the source line (Figure 3.2). It is therefore of some importance to consider what factors affect the broadening of the absorption line.

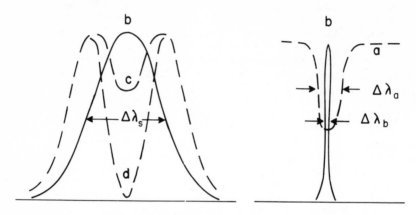

Figure 3.2. Absorption (a) and source (b) line profiles
(half-widths--Δλ).
^{c}Self-absorption. ^{d}Self-reversal.

Doppler broadening--due to the thermal activity of the atoms--predominates at low atomic concentrations. The Doppler half-width is given by:

$$\Delta\lambda_D = \frac{\lambda_c}{c} \cdot \left(\frac{8RT \ \ln Z}{m_a}\right)^{1/2} \qquad (3.4)$$

Here λ_c is the wavelength at the center of the line. It may be seen that Doppler broadening increases with temperature (T) and wavelength (λ), but decreases with the atomic weight (m_a) of the analyte. This broadening is symmetrical about λ_c, with a Gaussian profile.

Other factors which may have an effect upon the line profile are caused by pressure broadening due to atomic collisions with either like or foreign atoms. These effects do not yield a Gaussian-shaped profile (and may impart asymmetry) but are subordinate to the Doppler broadening at low atomic concentrations as noted. The total absorption line half-width in normal flames may be approximated as:

$$\Delta\lambda_a = [(\Delta\lambda_D)^2 + (\Delta\lambda_c)^2]^{1/2} \qquad (3.5)$$

and a compilation of these terms for a range of elements has been given by Winefordner *et al.*[7]

Atomic Absorption

For AA, the concept of the absorption coefficient (K_ν) is all important. It may be derived from first principles[8] by considering the passage of radiation through a defined isotropic layer. The ν subscript indicates that the coefficient is a function of the frequency ν or, of course, the wavelength. In practice, an integrated coefficient between ν and $\nu + d\nu$ can be defined:

$$\int_0^\infty K_\nu \, d\nu = \frac{h\nu}{c} \cdot B_o^j \cdot N_o \qquad (3.6)$$

If the atom is treated as a dipole oscillator (comprising two equal but oppositely charged poles) then f--the oscillator strength--may be defined:

$$f_o^j = \frac{m_e h\nu}{\pi e^2} \cdot B_o^j \qquad (3.7)$$

The f term is a measure of the average number of electrons per atom which may be excited, and characterizes the efficiency of the particular energy-level transition (only ground-state transitions are being considered in this simplified treatment). Combining Equations 3.6 and 3.7 gives:

$$\int_0^\infty K_\nu d\nu = \frac{\pi e^2}{m_e c} \cdot N_o \cdot f_o^j \qquad (3.8)$$

so that k_ν is a function of N_o and f only; there is no temperature term. Where available, f values are a useful guide to the utility of a particular transition (in practice, an analysis line). Some data are included in Table 6.1.

So far in this generalized discussion, it has
been shown that the practical spectral absorption
line has a finite width which depends upon the relative magnitude of the factors contributing to the
broadening. If it is accepted that the line width--
at common flame temperature--is controlled principally
by Doppler broadening, then a usable absorption
coefficient ($k_{\nu \cdot max}$) for absorption at the center of
the line may be defined thus:

$$K_{\nu \cdot max} = \frac{2}{\Delta \nu_D} \cdot \frac{\ell n 2}{\pi}^{1/2} \int_0^{\infty} k_{\nu} \, d\nu \qquad (3.9)$$

The $2(\ell n \, 2/\pi)^{1/2}$ term follows from treating the
Doppler Gaussian-shaped profile as approximating a
triangle (see Reference 9 for a more complete treatment and explanation). $\Delta \nu_D$ is the Doppler half-width
analogous to the $\Delta \lambda_D$ wavelength term defined in
Equation 3.4. From Equations 3.4 and 3.8, Equation
3.9 takes the form:

$$K_{\nu \cdot max} = \frac{e^2}{m_e} \cdot \left(\frac{m_a \pi}{2RT}\right)^{1/2} \cdot \frac{N_o f}{\nu_c} \qquad (3.10)$$

Thus, within the limitations of the various assumptions made, and for a given atom and a particular
transition (fixed ν or λ and f), the absorption
coefficient measured at the line peak is directly
related to N_o.

The absorption of energy by (ground-state) atoms
has been considered in this section, but nothing has
been said thus far about the relationship between
this derived absorption coefficient and the incident
light. The latter topic is considered later, when
it will be shown to conform, as might be expected,
to the classic Beer relationship.

The complexity of the above absorption coefficient
derivation stems in part from the need to define a
practical, measurable coefficient--in this case, a
coefficient for peak adsorption at the center of a
narrow line [assuming the line width to be controlled
by Doppler (Gaussian-shaped) broadening]. Other
coefficients may be similarly determined depending
upon the desired boundary conditions. Some simple
modifications to this basic expression will be
alluded to below.

Atomic Fluorescence

Atomic fluorescence is a result of resonance radiation absorption as illustrated by Equation 3.1, followed by re-emission at the same or some lower frequency:

$$R^{o*} \rightarrow R^o + h\nu' \qquad (3.11)$$

Not all the source energy is re-emitted, however, since some fraction is expended by various radiationless collisional (quenching) reactions. Some generalized transitions which yield the basic fluorescence radiations are shown in Figure 3.3.

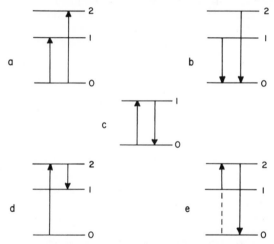

Figure 3.3. *Schematics of radiational transitions (solid arrows) between energy levels (0, 1, 2) for atomic spectrometry.*
a,b,c Resonance absorption, emission, fluorescence.
d Normal direct-line fluorescence.
e Thermally assisted direct-line fluorescence.

Resonance fluorescence--that is, absorption and re-emission of the same frequency radiation--is most often used for practical AF analysis, and only this type will be considered in the basic theoretical discussion given here. A fairly comprehensive review of the complex excitation and de-excitation processes of atoms in common flames has been given by, among others, Omenetto and Rossi.[10]

The spectral theory pertinent to AF as a viable spectrometric analytical procedure was given initially by Winefordner and Vickers,[9] and various fluorescence

intensity expressions--some considerably detailed--
have been developed in the literature since.   In
relating fluorescence emission to the source energy,
expressions for both the cell areas radiated and
viewed must be incorporated; some of the superficial
differences in derived AF intensity expressions stem
from the use of differing models and boundary conditions.   Winefordner *et al.*[11] stipulate incomplete
illumination of their model cell and subsequent
collection of a reduced fraction of the emitted
radiation at the monochromator slit.   The integrated
incident and fluorescence intensity parameters thus
derived are *per* unit solid angle.   It is expedient
here to assume the theoretical validity of a somewhat
simpler model, one in which the cell is totally
illuminated and the fluorescence emission in the
direction of the detector is measured.   A basic
expression for the (integrated) fluorescence energy
per unit area is then given by:

$$I_F = I_A \cdot \phi \cdot A_E \cdot Y \qquad (3.12)$$

A_E under these boundary conditions is an irradiated
area/emission area ratió, and Y allows for the fact
that, although the fluorescence is omnidirectional,
only some fraction is actually measured.   Y is expressed in steradians and usually takes the form
$\Omega/4\pi$ where Ω is the solid angle of fluorescence
emission "collected" by the measuring system.   The
remaining terms of Equation 3.12 are standard and
independent of the cell geometry.   Basically I_F is
directly related to the energy absorbed, and to ϕ--
the quantum efficiency (it must be reemphasized that
this simplified approach is for resonance fluorescence
only).
    The relationship between the measured fluorescence
intensity and N_O (that is, the cell atomic concentration) follows from the derivation of I_A; this is
briefly considered for both line and continuum
sources below.   For example, the following is derived
from Equations 3.24 and 3.25 for a line source at
low atomic concentrations:

$$I_A = Const. \ I_o \cdot N_o \cdot b \qquad (3.13)$$

so that from Equations 3.12 and 3.13, for given
conditions:

$$I_F = Const. \ I_o \cdot N_o \cdot b \cdot \phi \qquad (3.14)$$

This relationship is shown diagrammatically in Figure 3.4. At high atomic concentrations the linear relationship is lost. In general analytical terms it should be noted that the fluorescence signal is directly related to the incident radiation, so that analysis is optimized by utilizing as intense a source as possible.

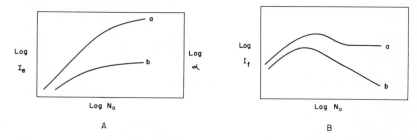

Figure 3.4. A. *Shape of growth curves for atomic absorption and emission.*
^{a}*AA line of source.*
^{b}*AE and AA continuum source.*
B. *Shapes of growth curves for atomic fluorescence.*
^{a}*Continuum source.*
^{b}*Line source.*

In the treatment developed above, it is shown also that I_F is dependent upon ϕ as defined. This spectrometric relationship--unique to atomic fluorescence--is of considerable importance (see References 12 and 13). In practical terms, an expression for the total fluorescence yield or quenching efficiency (y) is commonly incorporated. This may be defined as a *power* ratio, so that the cell geometry is not limiting:

$$y = \frac{P_{F\nu}}{P_{A\nu}} \tag{3.15}$$

Fluorescence quenching is a result of collisional, radiationless deactivation reactions during the finite lifetime (τ) of the excited atom. Hence, in addition to the transformations represented by Equation 3.11, and considering a flame-cell environment, the following generalized reaction must be taken into account in AF analysis:

$$R^{\circ}* + X \rightarrow R^{\circ} + X \tag{3.16}$$

The reaction kinetics for such de-excitation processes has been considered in general by, for example, Alkemade.[14] The treatment given here, specifically related to flames, follows the detailed derivative of Jenkins[15,16] using (pseudo) first-order rate constants. If r_R is the reaction rate of Equation 3.11--again considering only resonance fluorescence--and Σk_X [X] is the summation of Equation 3.16 for all the various X (flame) species, then:

$$y = \frac{r_R}{r_R + \Sigma k_X \text{ [X]}} \qquad (3.17)$$

(A self-absorption function has been considered to approximate unity in this and the above expressions; see Reference 11.) It is clear that the composition of the flame used for routine AF analysis is of considerable importance; [X] is commonly related to y by means of standard quenching cross sections.[15] Table 3.1 shows some y values for various flames and elements.[16] Further consideration of the relationship between fluorescence yield and the specific chemical flame species is given in Chapter 8. Alkemade[17] has derived expressions for nonresonance fluorescence yields.

Table 3.1

Two Examples of Calculated Fluorescence Yields (y)
in Various Flames   (Data from Reference 16)

Flame Composition	Temperature ($^\circ C$)	Fluorescence Yield Pb	Fluorescence Yield Tl
1) $2 H_2 - O_2 - 4 N_2$	1830	0.079	0.070
2) $6 H_2 - O_2 \ 4 N_2$	1530	0.100	0.099
3) $4 C_2 H_2 - O_2 - 4 N_2$	1930	0.067	0.042
4) $2 H_2 - O_2 - 10 Ar$	1530	0.220	0.330
5) $H_2 - O_2 - 4 N_2$	1330	0.069	0.051

One additional extension of this basic theory must be mentioned here. It has been shown[18] that lasers represent ultra-intense saturation sources. This means in effect that I_F has attained its maximum value, and also that the dependence on ϕ (or the y values as given in Table 3.1) is removed so that the fluorescence signal ceases to be a function of the particular flame chemistry used for atomization.

Continuum Sources

Theoretically, continuum sources may be used for both AA and AF; the basic form of the absorption equation is unchanged regardless of the source utilized, and to emphasize the similarities, these relationships are considered together in the following section (see also Reference 19).

AA sensitivity, when a continuum source is used, is controlled by the spectral band-pass of the instrument monochromator and, in general, compares unfavorably with the sensitivity obtainable using narrow line sources. The theoretical basis for this comparison was developed in the pioneering treatise on AA by Walsh[20] and it is popularly supposed that continuum sources for AA analysis are totally impractical. In several cases, however, it has been shown that the continuum source sensitivity is somewhat better than could be predicted by the limiting spectral theory considered here. For example, specific improvements have been noted[21] for atoms having complex spectra, suggesting the validity of absorption spread over several lines. The potential importance of spectral hyperfine structure has been emphasized by several people[22] and, in recent years, various attempts to determine absolute values for N_O have often shown discrepancies between continuum and line-source values which have been attributed[23] to such hyperfine structure unaccounted for in the source theory used.

For AF analysis, continuum sources are far more viable since the monochromator does not "see" the source. These spectral sources are frequently used as a matter of convenience for multielement analysis, but for most specific analytical applications the intensity available compares unfavorably with high-intensity line sources. Some practical applications of continuum sources are considered in Chapter 6.

Analytical Curves

The theoretical relationship between the measured AS radiation intensity parameter and the concentration of ground-state atoms in the absorption-emission cell must now be considered. It is instructive for the analyst to be reasonably familiar with this topic as a specific example of the fundamental relationship between the measured signal (×) and the atomic concentration--C of Equation 3.41. This relationship, in the form of practical analytical curves, is basic to all such analytical techniques.[24] The theoretical counterparts of the "real" analytical working curves are termed "growth curves." Only brief derivatives of growth curves for AA may be included here, but very similar reasoning would yield equivalent expressions for AE and AF. This cursory treatment follows the comprehensive work of Zeegers *et al.*,[25] which allows for both Doppler and collisional (Lorentz) line broadening, but most of the intermediate steps have of necessity been omitted, and the interested reader is urged to consult the cited references.

An expression[19] for the absorbed energy (integrated) *per* unit time and flame cross section is given by:

$$I_A = \int I_\lambda (1 - e^{-K_\lambda b}) \, d\lambda \qquad (3.18)$$

The special case (Figure 3.2) where a line source is used which is considerably narrower than the absorption line width should be considered further. An expression for a measurable absorption coefficient, based upon a line profile assumed due to pure Doppler broadening has been given above (Equation 3.10). To simplify the present discussion it is expedient to define a mean AA coefficient over the entire line width $(\overline{K})$, although it would, of course, be possible to replace this term by the $K_{\nu \cdot max}$ of Equation 3.10 by making the necessary transformations. (See Reference 25 for other mathematically simplifying constraints.) Under these conditions Equation 3.18 may be replaced by:

$$I_A = (1 - e^{-\overline{K}b}) \int_{\Delta\lambda_b} I_\lambda \, d\lambda \qquad (3.19)$$

and

$$\int_{\Delta\lambda_b} I_\lambda \, d\lambda = I_o \qquad (3.20$$

then

$$\frac{I_A}{I_o} = \alpha = (1 - e^{-\overline{K}b}) \qquad (3.21)$$

and

$$\overline{K}b = -\ln (1 - \alpha) \qquad (3.22)$$

Since

$$A = -\log (1 - \alpha) \qquad (3.23)$$

and

$$\overline{K} = f (N_o) $$

then

$$A = \text{Const. } N_o \cdot b \qquad (3.24)$$

This is a basic Beer-Lambert expression; a plot of absorbance (A) against N_o or, more practically, the atomic concentration (C) should be linear from this derivation. In fact, various processes act to cause bending of the curve at higher concentrations (as will be discussed briefly in Chapter 9) and at low concentrations, that is, for low optical density:

$$(1 - e^{-\overline{K}b}) \rightarrow \overline{K}b$$

so that Equation 3.21 may be simplified to:

$$\frac{I_A}{I_o} = \alpha = \overline{K}b \qquad (3.25)$$

and, from Equation 3.24:

$$\alpha = \text{Const. } N_o \cdot b \qquad (3.26)$$

(the constant depends upon the wavelength).

In the case of continuum sources, the absorption line width is considerably narrower than the spectral band width (s) of the monochromator, which in effect controls the absorption characteristics. If a constant $I_{o\lambda}$ may be assumed over the width s, then Equation 3.18 takes the form:

$$I_A = I_{o\lambda} \int_0^\infty (1-e^{-K_\lambda b}) \, d\lambda \qquad (3.27)$$

and

$$\alpha = \frac{I_A}{I_{o\lambda} \cdot s} \qquad (3.28)$$

Again, for the special case of low optical density:

$$(1-e^{-K_\lambda b}) \to K_\lambda b$$

and

$$\int_0^\infty K_\lambda \cdot d\lambda = f \, (b,\lambda,f,N_o)$$

so that

$$\alpha = \text{const. } N_o \qquad (3.29)$$

for a given λ, b and s, the plot of α against N_o (that is, per cent absorption *vs.* concentration) is linear also. The shapes of the resulting growth curves for AA are given as log-log intensity-N_o plots in Figure 3.4.[26] It may be seen that the line-source absorption curve has an extended linear range compared with the continuum.

As noted previously, Zeegers *et al.*[25] have also presented detailed derivations for AF (see also References 17 and 27) and AE growth curves, and it would be superfluous to present further duplication here; the relevant theory is closely similar to that outlined above for AA. The theoretical shapes of the final AF curves for both continuum and line sources are included also in Figure 3.4; neither curve is of much practical analytical use at high optical densities (concentrations). It must again be stressed that all the curves of Figure 3.4 are for idealized geometries and physical parameters. In practice, many factors act to cause deviation from these theoretical curves (see Chapter 9). As an example, one potentially disruptive factor often cited is that of resonance-type collisional line broadening, although it has also been suggested[5] that the importance of this particular effect is generally overestimated.

The Absorption Cell

In the previous sections, the concept of N_O as representing atomic concentration, and the relationship between various defined absorption coefficients and N_O have been discussed. What is in fact measured in AA, however, is the *fraction* of radiation absorbed (α) which has been defined as the ratio I_A/I_O. If the absorption process may be considered to occur within a definite cell of length b, then the general form of the Beer-Lambert law may be applied:

$$I = I_o e^{-Kb} \tag{3.30}$$

where I is the intensity after passage through the absorption cell, or:

$$\ell n \ (1-\alpha) = -Kb \tag{3.31}$$

The absorbance A, which is the parameter most familiar to atomic absorption analysts, is defined as:

$$A = \log \frac{I_o}{I} = -\log \ (1-\alpha)$$

$$= Const. \ N_o \cdot b$$

as given in Equation 3.24. It has also been noted above that A approximates α at low concentrations.

It is of course not possible to directly apply the Beer-Lambert relationship to flame absorption in the same way that it is applied to, for example, colorimetric analysis, since the flame is dynamic and an anisotropic medium and neither C nor b are constant for any extended time. However, within these limitations, the utmost care should be taken to devise the most favorable cell geometry, particularly with regard to alignment of the source beam within the flame and the relative dilutions of the sample within the various flames and burners. The dynamics of the system are also worth some considerations. For example, the residence time (τ_c) of an atom within an unenclosed flame cell (*i.e.*, within the source beam) is of the order of 10^{-4} seconds,[1] and efforts to increase flame temperatures may shorten this residence time even further. Some indication of this may be given by considering the flame velocities given in Tables 7.2 and 7.4. The only really major improvement in cell efficiency has come with the introduction of the enclosed cell devices considered in Chapter 7.

Comparison of AS Theoretical Radiation Intensities

A complete development[25] of the AS radiation intensity expressions (outlined briefly above for AA) gives a theoretical basis for a comparison of the individual AS techniques. Thus, the following *signal ratios* have been given by Winefordner and co-workers:[28,29]

$$\frac{\text{AA line source}}{\text{AF line source}} \sim \frac{\text{AA continuum source}}{\text{AF continuum source}} \sim \frac{1}{\phi Y} \qquad (3.32)$$

also

$$\frac{\text{AE}}{\text{AF continuum source}} \sim \frac{I_\lambda^B}{I_{o\lambda} \cdot \phi \cdot Y} \qquad (3.33)$$

and

$$\frac{\text{AE}}{\text{AF line source}} \sim \frac{I_\lambda^B \cdot \Delta\lambda_D}{I_o \cdot \phi \cdot Y} \qquad (3.34)$$

where I_λ^B is the intensity (spectral radiance) of a black body at peak wavelength λ (the other terms are as given previously).

Winefordner[28] has shown that the ratios of Equation 3.32 should be of the order of 10^3 for conventional flames. As a gross generalization, the signal ratios AF/AE and AA/AE should exceed unity for wavelengths below about 3000Å but AE should be theoretically superior for analysis lines greater than 4000Å.

Theoretical Detection Limits

The detection limit is defined (Chapter 9) not only by the relative intensity of the measured atomic spectrometric signal but by the noise level. Theoretical limiting-concentration expressions for all three AS methods have been derived by Winefordner,[30-32] and current practical compilations for various analytical arrangements are given later in Tables 3.5 and 3.6. Since it is demonstrably true that AA detection limits are not vastly superior to, for example, AF, as might be supposed from reasonable substitution into Equation 3.32, it appears that the noise levels--from all sources-- might be limiting for low levels of detection by AS. For both AF and AE the principal sources of electronic noise are shot noise (see Chapter 6) and background flicker noise. The former is usually of greater magnitude than the flicker noise and, where background emission is so low that shot noise is negligible, the detection limits are controlled by dark-current and amplifier noise. In AA analysis, source-flicker constitutes the principal noise source (as this is usually greater than all other background flicker noise), so that the detection limits for AA and AF are more closely comparable. With the use of nonflame cells, both background radiation and radiation flicker become vanishingly small, so that AF has an added advantage as a trace technique. Representative nonflame detection limits are given in Tables 7.5 through 7.10.

Dissociation and Ionization

So far this rapid review of the basic physics of AS analysis has omitted the most critical step, that is, the chemical conversion of the sample to the free atomic form. Table 3.2 is a representation of the relevant physicochemical reactions. A generalized metal compound dissociation reaction may be represented:

Table 3.2

Principal Physicochemical Transformations of Analyte R[a]

Instrument Component	Process	Phase	Physical State of Products	Chemical Form of Products
Sampling System	Nebulization	Liquid	Liquid-gas aerosol	$R^+ + A^-$
	Evaporation	Liquid	Supersaturate solution	
		Liquid & Solid, Solid & Vapor	Solid-gas aerosol	RA
Burner	Atomization	Vapor	Neutral nonexcited	R°, RA
	Ionization		Ionic nonexcited	R^+
	Absorption		Neutral excited	$R^{\circ*}$, RA*
			Ionic excited	R^{+*}
	Emission		Neutral nonexcited	R°, RA
			Ionic nonexcited	R^+
	Association		Neutral nonexcited	RA

[a] Vapor phase reactions without reference to time.

$$RA \overset{\rightarrow}{\leftarrow} R^\circ + A; \quad K_d = \frac{[R^\circ][A]}{[RA]} \tag{3.35}$$

so that a degree of dissociation term may be defined:

$$\alpha_d = \frac{[R^\circ]}{[R^\circ] + [RA]} \tag{3.36}$$

However, close estimations of either K_d or α_d as defined are next to impossible to obtain for the common emission-absorption cell environment. By making some gross assumptions, free energy calculations[33] may yield a first approximation but the chemical complexity of most samples and flames prohibits accurate data. Interferences attributable to incomplete dissociation reactions of the general class of "refractory compounds" are legion (Chapter 8). For this reason, it is not being too dogmatic to state that the atomization component of current flame AS equipment is less than satisfactory; this is a problem in addition to the problems associated with a variable geometric and dynamic cell mentioned above.

The forward reaction of Equation 3.35 is a function of the cell energy (temperature) available. This fact was well appreciated in the early days of AA analysis when innumerable interference problems were discovered to be associated with the use of low temperature flames chosen initially to promote a favorable N_j/N_o ratio. Figure 3.5(A) illustrates oxide dissociation trends for zinc and magnesium. A more complete appreciation of these problems has led over the years to the use of progressively hotter flames for routine AA analysis, since, in general, problems associated with incomplete or variable dissociation are potentially more troublesome than any advantages gained by optimizing the N_o term. For AE analysis, of course, high temperature chemical and plasma flames have long been used.

Although values may not be accurately assigned to the α_d term, one allied parameter that *can* be determined is the efficiency or degree of atomization (β) which may be defined:

$$\beta = \frac{N_o}{N_{total}} \tag{3.37}$$

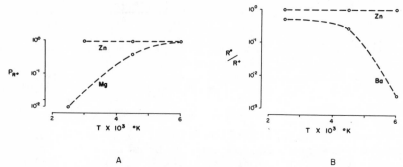

Figure 3.5. Examples of the control by:

  A. Dissociation (P_{RO}--partial pressure of metal
     based solely upon oxide dissociation), and

  B. Ionization (ratio $R°/R^+$) processes of atomic
     absorption and emission. (Data from Reference
     33.)

where N_{total} is the sum of the concentrations of all
R species in the flame cell--atoms, compounds and
ions. This useful and practical coefficient is
considered again in Chapter 7.

One of the foremost problems encountered in AS
analysis is that of the formation of metal oxides
in the atomization cell, and this difficulty is
especially relevant to heavy-metal analysis. The
high oxygen partial pressure in certain zones of
air/oxygen flames favors the reverse reaction of
Equation 3.35, correspondingly depleting the free-
metal atom population. This ubiquitous phenomenon
(see Reference 14 for a more detailed discussion)
has led to considerable research on flame structures
and chemistry, to the introduction of organic solvents, fuel-rich and separated flames, and to close
control of the flame zone used for analysis. These
topics are of considerable importance to heavy-metal
analysis and are considered in detail in Chapter 7.
Table 3.3 shows dissociation energies for various
RO compounds.

The oxide dissociation discussed above is
essentially a problem of compound formation within
the flame. It is important to remember, however,
that not all chemical species of a metal likely to
be encountered in natural waters will be dissociated
with equal facility, and special problems may exist

Table 3.3

Ionization Energies (E_i) and Dissociation Energies $(E_D \cdot _{Ox})$
of Oxides (RO)

Element	E_i (ev)	$E_D \cdot _{Ox}$ (ev)	Element	E_i (ev)	$E_D \cdot _{Ox}$ (ev)
Ag	7.57	1.4	Os	8.73	
As	9.81	4.9	Pb	7.42	4.1
Au	9.22		Pd	8.33	
Bi	7.29	4.0	Pt	9.00	
Cd	8.99	3.8	Re	7.87	
Co	7.86		Rh	7.45	
Cr	6.76	4.2	Ru	7.34	
Cu	7.72	4.9	Sb	8.64	3.2
Fe	7.87	4.0	Se	9.75	3.5
Ga	6.00	2.5	Sn	7.34	5.7
Ge	7.88	6.5	Ta	7.88	
Hf	6.80		Te	9.01	4.0
Hg	10.43		Ti	6.82	6.9
In	5.78	1.1	Tl	6.11	
Ir	9.00		V	6.74	6.4
Mn	7.43	4.0	W	7.98	
Mo	7.10	5.0	Zn	9.39	4.0
Nb	6.88	4.0	Zr	6.84	7.8
Ni	7.61				

where metals are introduced into the flame cell as
extracted chelates. A documented example of such
difficulties encountered with mercury analysis is
noted in Chapter 10. It is of utmost importance to
remember that while AS analysis remains a comparative
analytical technique the standards used must be
carefully matched; this necessitates at least a
reasonable familiarity with the sample chemistry.

With the use of increasingly higher flame-cell
temperatures, ionization of the analyte atom is
promoted:

$$R° \rightleftarrows R^{n+} + ne \tag{3.38}$$

At equilibrium an ionization constant (K_i) may be
defined in the usual fashion:

$$K_i = \frac{[R^{n+}][e]}{[R°]} \tag{3.39}$$

and, analogous to Equation 3.36, the degree of
ionization is:

$$\alpha_i = \frac{[R^{n+}]}{[R^{n+}] + [R^\circ]} \qquad (3.40)$$

K_i is directly related to the cell energy (temperature)
as given by the Saha equation (see Reference 8), so
that increased ionization of the sample atoms is a
potentially troublesome problem in AA and AF where
high temperature flames are used.[34] However, it may
be seen from Table 3.3 that the ionization energies
(E_i in ev) for the heavy metals under consideration
here are high; ionization of these elements is not
likely to cause problems with conventional flames.
Figure 3.5(B), for example, illustrates R°/R^+ with
temperature for zinc as compared with a more readily
ionized alkali earth metal; zinc is not appreciably
ionized over this temperature range. Table 3.4
lists degrees of ionization for the heavy metals
calculated[35] for air acetylene and nitrous oxide-
acetylene flame temperatures. In AE analysis, where
very high temperature chemical and plasma flames are
used, appreciable ionization is to be expected, and
the potential use of ion analysis lines should not
be overlooked.

Absolute Analysis

Any quantitative analytical procedure depends
upon a relationship of the type:

$$\times = f\ (c) \qquad (3.41)$$

where $\times$ is the measured parameter and C the concentration (see Reference 36). The theoretical derivations of expressions for I_A, I_O and I_F in terms of
N_O have been discussed above. Hence, if all the
relevant parameters were known, it would be possible
to directly determine sample concentrations (or, at
least, ground-state atomic populations). As presently
utilized, of course, all practical AS techniques
determine concentrations relative only to standard
solutions; the limitations this imposes upon the
accuracy of the final data are noted in Chapter 9.
Rann[37] initially considered the major problems--
for example, dissociation, pressure broadening,
characterization of the emission lines--likely to

Table 3.4

Degree of Ionization (%) Computed for Two Standard
Flame Temperatures[a] (Data from Reference 35)

Element	Air-Acetylene (2280°C)	N₂O-Acetylene (2930°C)	Element	Air-Acetylene (2280°C)	N₂O Acetylene (2930°C)
Ag		1.6	Pb	0.1	4.0
As			Pd		1.2
Au			Pt		0.1
Bi		0.5	Rh		1.3
Cd		0.2	Ru		1.4
Co		1.3	Sb		0.3
Cr	0.3	8.5	Se		
Cu		1.2	Sn	0.1	3.0
Fe		1.6	Te		0.1
Ga	3.9	17.6	Ti	0.4	12.2
Hg			Tl	2.1	21.8
In	5.1	29.2	V	0.3	9.7
Mn	0.1	3.1	W		0.7
Mo	0.1	2.9	Y	0.5	15.2
Ni		1.1	Zn		0.1
Os		0.5	Zr	0.2	7.6

[a] Where no values are given, ionization < 0.1% under the assumption used for the computation; see Reference 35.

mitigate against absolute AS analysis. In recent years, several people (see Chapter 7) have, in fact, calculated N_O values and compared these with sample concentrations to demonstrate the atomization efficiency of various flame combustion mixtures, so it appears that the major obstacles to direct analysis are chemical. These latter experiments have utilized continuum source absorption techniques.[37] It is likely to be some time before this type of work is developed sufficiently to be a practical alternative to the present comparative methods.

GENERAL CHARACTERISTICS

System Components

The component parts of an idealized atomic
spectrometric analysis system are shown in Figure
3.6. Not all commercially available instruments
incorporate all the features illustrated, but most
basic equipment may be added to or modified as required. A major point to note is that all AS
techniques are closely allied and that the same
instrumentation may, with slight modification, be
used for all three. The source radiation required
for AA and AF may be identical and one detection-
readout system serves for all. The heart of the AS
system is the sampling-atomization device, and this--
for reasons discussed in various sections of this
volume--is probably the least efficient part of the
instrument. This component must be optimized for
each analytical use as discussed in Chapter 7. Many
of the individual parts of the system illustrated
in Figure 3.6 will probably be incorporated into a
single commercial package, but the analyst must be
on guard to purchase equipment which provides
maximum space and flexibility for the atomization
unit. The various spectral sources available and
basic detection components are considered in
Chapter 6.

Samples and Sampling

Water Analysis

Atomic spectrometry is basically a technique
for solution analysis, and is hence particularly
suited to samples that are liquid in the natural
state, or in those cases where mandatory pre-analysis
chemistry yields a final liquid phase. It would
therefore appear, *a priori*, that atomic spectrometry
should be a preferred choice for natural water
analysis. But in fact, the exponential growth of
AA analysis in particular has been largely in the
fields of biology (agriculture, clinical, etc.),
geochemistry (assaying etc.) and metallurgy--pre-
dominantly areas where the raw sample is not
initially in liquid form. A rapid review of the
application texts[38] well illustrates the paucity of
applications to natural water analysis. This bias

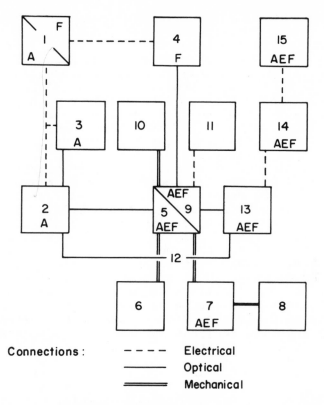

Connections :   – – – –   Electrical
————   Optical
====   Mechanical

*Figure 3.6.   Schematic of idealized atomic spectrometric
analysis system.*
1. *AA/AF radiation source power supply and source
   modulator.*
2. *AA radiation source.*
3. *Continuum reference source.*
4. *AF radiation source.*
5. *Burner/plasma AE source, burner/plasma AA/AF
   sample atomization, Hg absorption tube.*
6. *Burner/plasma gas supplier.*
7. *AA/AE/AF nebulization for continuous sample
   introduction.*
8. *Automated sample introduction.*
9. *Flameless AA/AF atomization device, flameless
   AE source.*
10. *Controlled-atmosphere gas supply.*
11. *Power supply for flameless atomization device.*
12. *Reference beam.*
13. *Monochromator/resonance detector.*
14. *AA/AF signal demodulator, detection and
    amplification.*
15. *Analog/digital display.*

has even led to several attempts to devise practical
methods for the direct analysis of solid samples.
Only one such device[39] has been marketed commercially,
and this technique has been less than successful.
Flame analysis invariably requires the nebulization
of liquids as the only practical means of obtaining
reproducible atomic vapor distributions within the
analysis cell. Nonflame devices also currently
utilize liquid samples for much the same reasons,
but the future potentialities here for direct solid
atomization are largely unexplored.

Application of AS to natural water analysis has
been limited principally because, for trace-metal
analysis and the use of continuous nebulization,
the detection limits available are generally insufficient. As a gross generalization, "continuous
sampling mode" flame analysis gives usable data to
a lower limit of around 1 mg/ℓ (ppm), or, at best,
to 0.1 mg/ℓ. Values by this analysis mode are, of
course, comparative not absolute. Since detections
of around 1 μg/ℓ (ppb) are required, pre-analysis
sample concentration is mandatory,[40] and the initial
advantage of direct solution analysis is negated.
(For major cation analysis--not considered here--the
sensitivity of AS is frequently insufficient to
detect the small variations frequently sought.)
Within the last few years this situation has been
radically changed by the introduction of nonflame
(absolute) AS methods of greatly superior detection
capability.

Field Use

AS equipment is robust enough and sufficiently
portable for direct field use. For shipboard
operation[41] the full potentialities of the equipment
need not be impaired so long as the optical train is
protected against vibrations and the flame-cell
geometry does not fluctuate too wildly. Some additional voltage stabilization may also be required.
It should also be possible to attain a high degree
of portability for small lake and river analysis
in situ; advantage should be taken of the liquid
organic fuels (noted in Chapter 7).

Automation

Methods to speed analysis or to free the operator
for other work are always at a premium.[42] Using the

continuous sampling mode, it is a very simple matter
to couple an AA nebulizer to an automatic rotating
sample-changer,[40] but the utility of these methods
depends upon having routine samples within a sufficiently high concentration range. Some thought
has also been given to simultaneous multielement
analysis by AA. Very early developmental work was
done by Dawson and co-workers,[43] and Walsh[44] has
described two AA multielement procedures using,
respectively, resonance detection and selective
modulation (see Chapter 6). A multichannel
spectrometer suitable for AE analysis has also been
described.[45]

The only commercially available multielement
analysis system at present is the Technicon AFS 6
spectrometer,[46] which uses sequentially pulsed, high
intensity (hollow cathode) lamps and a time-phased
detection system (see also Reference 44 for a
similar arrangement). These AF instruments are
ideal for use where a large number of closely identical samples need to be analyzed for several metals
within well-defined concentration ranges (soil extracts, for example[47]). For natural water analysis,
however, one is more often concerned with continuous
variations in the metal concentrations in time as
well as in space. For these purposes an automated
continuous monitoring spectrometer would be highly
desirable. The continuous sampling operation of
flame AS theoretically lends itself to this type of
operation,[48] and the Technicon instrument offers a
practical capability analogous to continuous
colorimetric AutoAnalyser techniques.

Limits and Limitations

The particular advantages and some of the more
undesirable features of the common trace-metal
analysis techniques have been considered in the previous chapter. It remains now to briefly outline
the operating boundaries of AS in a little more
detail. One AS feature in particular--the direct
solution analysis capability--has been cited as
particularly advantageous. But the water analyst
may well have to look at various coexisting solid
phases (a highly recommended routine for heavy-metal
evaluations), so that use of other techniques noted
previously might well offer theoretical advantages
at times. In practice, however, it is likely that
a wide range of equipment will not be readily
available; and there is a certain inertia which

mitigates against frequent fluctuation between analytical methods. In any case, most laboratories will currently be utilizing pre-analysis treatment of the water samples, so that the addition of an extraction or dissolution routine will not be too disruptive. In all cases, the choice, or otherwise, of one of the AS methods will be dictated by a multitude of factors peculiar to the individual laboratory.

Universality

Almost all the metals included in Figure 2.1 may theoretically be determined by all three AS techniques, but the choice in any given situation will depend upon considerations such as the sample matrix and concentration range of the test element. Figure 3.7(a) shows the metals determinable in concentrations less than 100 µg/ℓ under ideal conditions. It is considered that nonflame atomization techniques will predominate in the field of natural water analysis in the not-too-distant future; Figures 3.7(b) and 3.7(c) give elements suitable for carbon-filament analysis (both AA and AF) and for AA furnace detection, respectively (both taking 100 µg/ℓ as an arbitrary limit). These illustrative examples purposely refer to commercially available accessories, and the values have been standardized to relative concentration units.

Selectivity

The selectivity of the chosen method refers to its ability to determine an element in the presence of many elements. This is an exceedingly important concept for most working analysts. Both AA and AF are very superior tools in this respect. AE is generally considerably less so because of the need to separate analysis lines from concomitant emissions. Spectral interferences are considered in Chapter 8.

Detection Limits and Sensitivity

Published compilations of detection limits offer a valuable guide to the suitability of marrying a specific analysis to a specific problem; but they should serve *only* as a guide. Uncritical reliance on these data is fraught with danger for the unwary.

Ti	V	Cr	Mn	Fe	Co	Ni	Cu	Zn	Ga	Ge	As	Se
Zr	Nb	Mo	Tc	Ru	Rh	Pd	Ag	Cd	In	Sn	Sb	Te
Hf	Ta	W	Re	Os	Ir	Pt	Au	Hg	Tl	Pb	Bi	**a**

Ti	V	Cr	Mn	Fe	Co	Ni	Cu	Zn	Ga	Ge	As	Se
Zr	Nb	Mo	Tc	Ru	Rh	Pd	Ag	Cd	In	Sn	Sb	Te
Hf	Ta	W	Re	Os	Ir	Pt	Au	Hg	Tl	Pb	Bi	**b**

Ti	V	Cr	Mn	Fe	Co	Ni	Cu	Zn	Ga	Ge	As	Se
Zr	Nb	Mo	Tc	Ru	Rh	Pd	Ag	Cd	In	Sn	Sb	Te
Hf	Ta	W	Re	Os	Ir	Pt	Au	Hg	Tl	Pb	Bi	**c**

◯ Absorption ◹ Fluorescence ◺ Emission

Figure 3.7. Partial periodic tables showing elements determinable
in concentrations less than 100 ppb.
a. Continuous (flame) AA/AE/AF analysis.
b. Filament (discrete, flameless) AA/AF analysis
(N.B. limited data currently available for AF).
c. Furnace AA analysis (discrete, flameless,
commercially available).

The only advantage to such listings--apart from
indicating whether the element is determinable at
all--is that some indication is given of how close
to the limiting detection of the method the analysis
may be. This gives some rough estimate of the expected precision. The practiced analyst is far more
concerned with problems of selectivity and matrix,
with sensitivity, and probably with analytical speed,
availability of equipment and similar practical
problems. Theoretical aspects of the detection of
limiting concentrations has been considered above
and will be looked at again from a more practical
standpoint in Chapter 9. Tables 3.5 and 3.6 show
some recent literature detection limits for continuous
sampling mode flame AS. It is of the greatest importance to appreciate that these data have invariably
been obtained under ideal conditions. Detection
limits obtained for the variety of nonflame atomization cells are not directly comparable and are hence
not included here; some tabulations are included in
Chapter 7 where these atomizers are considered in
more detail. The latter values are given in terms
of absolute concentrations so that, to afford some
rough comparison with the conventional flame limits
of Tables 3.5 and 3.6, the analyst must take account
of the sample volume capacity of each module.
    The sensitivity of an analytical procedure is
concerned with the ability to resolve small differences between quantities of the analyte. In effect,
"sensitivity" may thus be equated with the slope of
the analytical working curve (see Chapter 9) and
thus is a useful, practical parameter delineating
the utility of a particular technique, although
general use of this term has been virtually confined
to AA. Sensitivities (Tables 6.3 and 9.2) are unique
to each analytical line and to the concentration too
if the working curve is nonlinear. In the past,
"sensitivity" has been used interchangeably with
"detection limit," though this practice is now rare,
understandably so considering the inevitable
confusion (but see Reference 49).

Accuracy and Precision

    Statistical reporting of the analytical data is
discussed in Chapter 9. Both random and nonrandom
error may be introduced at any period from the sample
collection stage to the final analysis. It is fair
to say that the actual instrumental analysis step
usually contributes the least problems in this area,

Table 3.5

Atomic Absorption Detection Limits:
Continuous Sampling Mode (µg/ℓ)

Element	Line (Å)	Chemical Flame	Plasma	Absorption Tube	Nonflame
Ag	3281	0.5[50]	500[51a]		100[52b]
As	1937	100[53]			
Au	2428	10[54]			
Bi	2231	40[53]		100[55]	
Cd	2288	0.6[56]		0.4[57]	1.8[58b]
Co	2407	5[38]		13[57]	
Cr	3579	3[53]			
Cu	3248	2[53]	1,500[51a]	5[55]	3[58b]
Fe	2483	5[29]			
Ga	2874	70[59]			
Ge	2652	100[54]			
Hf	3073	8,000[53]			
Hg	2537	200[54]		500[55]	
In	3039	50[38]			
Ir	2640	2,000[53]			
Mn	2795	2[53]		5[55]	2[58b]
Mo	3133	100[60]			
Nb	3349	1,000[53]			
Ni	2320	5[38]		10[55]	
Os	2909	600[61]			
Pb	2833	10[38]		20[55]	
Pd	2476	20[38]			
Pt	2659	100[38]			
Re	3460	800[62]	30,000[63a]		
Rh	3435	30[38]			
Ru	3499	300[38]			
Sb	2176	70[54]		100[55]	
Se	1960	100[50]			
Sn	2246	30[50]			
	2863			50[64]	
Ta	2715	2,000[53]			
Te	2143	100[54]		20[55]	
Ti	3643	35[60]			
	4668		5,000[63a]		
Tl	2768	200[38]		50[55]	
V	3184	40[60]	2,000[63a]		
	4660		3,000[63a]		600[65c]
Zn	2139	2[38]	0.6[57a]		
Zr	3601	240[60]			

[a]r.f. induction coupled.
[b]Furnace.
[c]Arc.

Table 3.6

Atomic Fluorescence and Emission Detection Limits:
Continuous Sampling Mode (μg/ℓ)

Element	Line (Å)	Fluorescence Line Source	Fluorescence Continuum Source	Emission Chemical Flame	Plasma
Ag	3281	0.1[66]	1[67]	20[68]	1,000[69a]
As	1937	100[69]			
	2288				100[70a]
	2350			2,000[71]	
Au	2676	5[72]	3,500[73]	500[68]	20[74a]
Bi	2231	5[75]		3,000[71]	
	3024	100[69]			
	3068		2,000[73]		2,000[74a]
Cd	2288	0.003[29]	80[73]		2[76b]
	3261			2,000[68]	
Co	2407	5[77]	500[78]		
	3454			50[68]	3[70a]
Cr	3572	50[69]			
	3579		10,000[79]		1[70a]
	4254			5[68]	
Cu	3248	1[79]	350[73]	10[29]	20[80b]
	3274			10[68]	
Fe	2483	8[69]	1,000[78]	50[29]	
	3720			50[68]	5[70a]
Ga	4172	10[75]	5,000[79]	10[68]	150[81b]
	4680		20,000[73]		
Ge	2652	100[82]		500[68]	
Hf	2866	3,000[66]			
	3400				10[70a]
Hg	2537	0.2[83]		700[71]	
In	4105	200[79]	2,000[79]		
	4511	100[66]		5[68]	300[81b]
Ir	2544		100[79]		
Mn	2795	1[84]	150[85]		
	4031			5[68]	150[81b]
Mo	3133	460[60]		160[60]	800[51a]
	3194			300[86]	
	3798			30[87]	
Nb	4059			60[86]	10[70a]
Ni	2320	3[79]	3,000[79]		
	3415		5,000[88]	30[68]	
	3525				6[70a]
Pb	4058	10[89]	3,000[79]	200[68]	8[70a]
Pd	3405	40[90]	50,000[79]		400[81b]
	3635			50[68]	
Pt	3064				100[74b]

Table 3.6, continued

Element	Line (Å)	Fluorescence		Emission	
		Line Source	Continuum Source	Chemical Flame	Plasma
Re	3461			300[87]	
	4889			1,500[91]	
Rh	3692	3,000[79]	10,000[79]	300[92]	50[81b]
Ru	3728			300[92]	
Sb	2311	50[79]			
	2598			200[71]	200[70a]
Se	1960	40[75]			
	2040			3,000[71]	
Sn	2430			500[71]	
	2840			300[68]	
	3034	50[93]			200[74b]
Te	2143	5[75]		600[71]	
	2386			2,000[94]	
Ti	3200	4,000[95]			
	3349				3[70a]
	3656			500[86]	
	3999			200[86]	
Tl	3776	8[66]	500[85]	90[92]	
V	3184	70[60]		40[60]	
	4112			2,000[71]	
	4379			50[86]	6[70a]
W	4009			400[86]	
	4302				1,000[80b]
Zn	2139	0.02[83]	30[78]	50,000[29]	9[70a]
Zr	3438				5[70a]
	3520			1,200[86]	

[a] r.f. induction coupled.

[b] h.f. (microwave) plasma.

and the magnitude of errors introduced at this stage may be determined with little effort. Most trace-metal techniques--not only AS--are frequently operated close to the operating limits for natural water analysis; the precision of much of the literature data is poor but this does not, of course, negate use of these values so long as the limitations are well understood and reported. Accuracy is potentially even less reliable in many cases than precision, particularly in the absence of reliable standards. One published comparative study of relative accuracy and precision for several aqueous trace-metal techniques has been given in Table 2.14.

Specific Disadvantages

The use of AS methods involves several potentially major disadvantages whose varying importance depends predominantly upon the use made of the data. For example, these methods yield *total* values and no chemical speciation is possible except by pre-analysis manipulation. In addition, the methods are destructive and comparative, so that values obtained can be only as good as the standards used. Interference problems as a function of the total matrix are considered in Chapter 8.

Practical Comparison of
AS Techniques

The total range of trace elements which may be determined by AS methods (Figure 3.7) is very wide (the theoretical relationship between analysis line wavelength and the particular suitability of AE, AA, or AF has been briefly noted above). However, in spite of the remarkable resurgence of AE technique in recent years, only a few metals are likely to be amenable to AE analysis unless the analyst has specialized equipment (plasma atomization, for example). The initial eclipse of AE by AA and AF was, to a large extent, due to some real advantages of the latter in terms of simplicity of operation, but it is also a fact that much of the supposed superiority of AA is fictitious in many cases. This topic has been well documented,[1] and the principal culprit is a natural tendency to generalize the characteristics of each mode. Each sample type is unique in some respect and may well exhibit characteristics which coupled with some experience on the part of the analyst, will suggest the use of a particular technique. It is hoped that helpful guidelines have been given both here and in the cited references, but these should be considered as just that, and not as dogmatic ground rules. It may well be true, as Robinson[96] has said, that "the average individual laboratory using commercial equipment and the 'average' analytical chemist can expect to get more reliable quantitative results using atomic absorption on a day-to-day basis"; but many natural water samples will demand a degree of analytical skill somewhat above "average." The major disadvantage of AE is that it is not amenable to the nonflame atomization operation currently available to AA and AF that will permit direct analysis of many water samples.

AF analysis has, to date, been sparsely used, in spite of the fact that the same basic equipment is common to both AA and AF. Some practical advantages of AF are:

1. The intensity of fluorescence emission may be increased directly by increasing the source intensity.
2. Continuum sources may be used.
3. Flame AF requires simpler burners and combustion mixtures than does AA.
4. AF (and AE) signals are direct deviations from the background, whereas the measured AA emission is a difference between two relatively intense signals. In the former case, electronic amplification is a more rewarding procedure, and photon-counting detection can be used to advantage.

The Literature

Most of the standard texts are referenced frequently in this volume. For flame emission the works of Herrmann and Alkemade,[97] Dean[98] and Mavrodineanu and Boiteux[99] are still standard though somewhat dated. Considerably more texts have been published specifically on AA, but these are all too frequently predominantly "recipe" manuals. The best single volume to date is that by Rubeska and Moldan.[8] The first two volumes of the treatise edited by Dean and Rains[100,101] is unrivaled for most theoretical aspects of both AE and AA, but the coverage of non-flame devices and AF is minimal, and no other single work specifically covers either of these latter topics. However, Winefordner and co-workers have published several comprehensive reviews of AF.[29] The texts which briefly mention the application of AS to water analysis are too numerous to catalog; many are referenced throughout this volume. Mavrodineanu has prepared two bibliographies[102,103] which comprehensively cover all analytical aspects of flame spectroscopy over the periods 1800-1966 and 1966-68 respectively. Later additions may be obtained from the manufacturers of AS equipment. For a theoretical background to aqueous chemistry, the book by Stumm and Morgan[104] has no peer.

REFERENCES

1. Pickett, E. E. and S. R. Koirtyohann. *Anal. Chem., 41,* 28A (1969).
2. Boumans, P. W. J. M. *Spectrochim. Acta, 23B,* 559 (1968).
3. West, T. S. in *U.S. National Bureau of Standards Monograph 100* (W. W. Meinke and B. F. Scribner, eds.), U.S. Gov't. Printing Office, Washington, D.C., 1967, pp. 215-301.
4. Lockyer, R. *Advan. Anal. Chem. Instru., 3,* 1 (1964).
5. Alkemade, C. T. J. *Appl. Opt., 7,* 1261 (1968).
6. Cresser, M. S., P. N. Keliher and C. Wohlers. *Spect. Lett., 3,* 179 (1970).
7. Winefordner, J. D. and J. M. Mansfield. *Appl. Spectr. Rev., 1,* 1 (1967).
8. Rubeska, J. and B. Moldan. *Atomic Absorption Spectrophotometry* (English translation, P. T. Woods, ed.), Iliffe, London, 1969.
9. Winefordner, J. D. and T. J. Vickers. *Anal. Chem., 36,* 161 (1964).
10. Omenetto, N. and G. Rossi. *Spectrochim. Acta, 25B,* 297 (1970).
11. Winefordner, J. D., M. L. Parsons, J. M. Mansfield and W. J. McCarthy. *Anal. Chem., 39,* 436 (1967).
12. McCarthy, W. J. and J. D. Winefordner. *Spectrochim. Acta, 23B,* 25 (1967).
13. Pearce, S. J., L. de Galan and J. D. Winefordner. *Spectrochim. Acta, 23B,* 793 (1968).
14. Alkemade, C. T. J. in *Flame Emission and Atomic Absorption Spectrometry,* Vol. 1 (J. S. Dean and T. C. Rains, eds.), Marcel Dekker, New York, 1969, pp. 101-150.
15. Jenkins, D. R. *Spectrochim. Acta, 23B,* 167 (1967).
16. Jenkins, D. R. *Spectrochim. Acta, 25B,* 47 (1970).
17. Alkemade, C. T. J. in *Atomic Absorption Spectroscopy* (R. N. Dagnall and G. F. Kirkbright, eds.), Butterworths, London, 1970, pp. 73-98.
18. Piepmeier, E. H. *Spectrochim. Acta, 27B,* 431 (1972).
19. de Galan, L., W. W. McGee and J. D. Winefordner. *Anal. Chim. Acta, 37,* 436 (1967).
20. Walsh, A. *Spectrochim. Acta, 7,* 108 (1955).
21. Butler, L. R. P. and J. A. Brink. in *Flame Emission and Atomic Absorption Spectrometry,* Vol. 2 (J. A. Dean and T. C. Rains, eds.), Marcel Dekker, New York, 1971, pp. 21-56.
22. Allan, J. E. *Spectrochim. Acta, 24B,* 13 (1969).
23. Willis, J. B. *Spectrochim. Acta, 26B,* 177 (1971).
24. de Galan, L. *Spect. Lett., 3,* 123 (1970).
25. Zeegers, P. J. T., R. Smith and J. D. Winefordner. *Anal. Chem., 40,* 26A (1968).
26. Winefordner, J. D. and R. C. Elser. *Anal. Chem., 43,* 24A (1971).
27. Hooymayer, H. P. *Spectrochim. Acta, 23B,* 567 (1968).

28. Winefordner, J. D. in *Atomic Absorption Spectroscopy* (R. N. Dagnall and C. F. Kirkbright, eds.), Butterworths, London, 1970, pp. 35-50.

29. Winefordner, J. D., V. Svoboda and L. J. Cline. *CRC Crit. Rev. Anal. Chem.*, *1*, 233 (1970).

30. Winefordner, J. D. and T. J. Vickers. *Anal. Chem.*, *36*, 1947 (1964).

31. Winefordner, J. D. and T. J. Vickers. *Anal. Chem.*, *36*, 1939 (1964).

32. Winefordner, J. D., M. L. Parsons, J. M. Mansfield and W. J. McCarthy. *Spectrochim. Acta*, *23B*, 37 (1967).

33. Addink, N. W. H. in *U. S. National Bureau of Standards Monography 100* (W. W. Meinke and B. F. Scribner, eds.), U.S. Gov't. Printing Office, Washington, D.C., 1967, pp. 121-148.

34. Manning, D. C. and L. Capacho-Delgado. *Anal. Chim. Acta*, *36*, 312 (1966).

35. Woodward, C. *Spect. Lett.*, *4(6)*, 191 (1971).

36. de Galan, L. and G. F. Samaey. *Spectrochim. Acta*, *25B*, 245 (1970).

37. de Galan, L. and G. F. Samaey. *Anal. Chim. Acta*, *50*, 39 (1970).

38. Slavin, W. *Atomic Absorption Spectroscopy*, John Wiley and Sons, New York, 1968.

39. Venghiattis, A. A. *Spectrochim. Acta*, *23B*, 67 (1967).

40. Burrell, D. C. *Anal. Chim. Acta*, *38*, 447 (1967).

41. Burrell, D. C. Proceed. 3rd Int. Atom. Fluoresc. Spectr. Cong., 1972, pp. 409-428.

42. Gafford, R. D. Proc. of FAO Tech. Conf. on Marine Poll., 1970.

43. Dawson, J. B., D. J. Ellis and R. Milner. *Spectrochim. Acta*, *23B*, 695 (1968).

44. Walsh, A. in *Atomic Absorption Spectroscopy* (R. N. Dagnall and C. F. Kirkbright, eds.), Butterworths, London, 1970, pp. 1-10.

45. Mavrodineanu, R. and R. C. Hughes. *Appl. Opt.*, *7*, 1281 (1968).

46. Mitchell, D. G. and A. Johansson. *Spectrochim. Acta*, *25B*, 175 (1970).

47. Dagnall, R. N., G. F. Kirkbright, T. S. West and R. Wood. *Analyt. Chem.*, *43*, 1765 (1971).

48. Burrell, D. C. *Atomic Absorption Newsletter*, *7*, 65 (1968).

49. Donega, H. M. and T. E. Burgess. *Anal. Chem.*, *42*, 1521 (1970).

50. Kahn, H. L. *Advan. Chem. Ser.*, *73*, 183 (1968).

51. Veillon, C. and M. Margoshes. *Spectrochim. Acta*, *23B*, 503 (1968).

52. Woodriff, R. and R. W. Stone. *Appl. Opt.*, *7*, 1337 (1968).

53. Perkin-Elmer Corp., P. E. 403 Operating Manual, 1971.

54. Varian-Techtron Pty Ltd., Techtron AA5 Operations Manual, 1971.
55. Koirtyohann, S. R. and C. Feldman. *Develop. Appl. Spectr.,* *3,* 180 (1964).
56. Koirtyohann, S. R. *Develop. Appl. Spectr., 6,* 67 (1968).
57. Fuwa, K. and B. L. Vallee. *Anal. Chem., 35,* 942 (1963).
58. Woodriff, R. and G. Ramelow. *Spectrochim. Acta, 23B,* 605 (1968).
59. Slavin, W. *Appl. Spectrosc., 20,* 281 (1960).
60. Dagnall, R. M. and G. F. Kirkbright. *Anal. Chem., 42,* 1029 (1970).
61. Fernandez, F. *Atomic Absorption Newsletter, 8,* 90 (1969).
62. Fiorino, J. A., R. M. Kniseley and V. A. Fassel. *Spectrochim. Acta, 223,* 413 (1968).
63. Wendt, R. H. and V. A. Fassel. *Anal. Chem., 38,* 337 (1966).
64. Agazzi, E. J. *Anal. Chem., 37,* 304 (1965).
65. Marinkovic, M. and T. J. Vickers. *Appl. Spectrosc., 25,* 319 (1971).
66. Zacha, K. E., M. P. Bratzel, J. D. Winefordner and J. M. Mansfield. *Anal. Chem., 40,* 1733 (1968).
67. Winefordner, J. D. and J. M. Mansfield. *Fluorescence: Theory, Instrumentation and Practice,* Marcel Dekker, New York, 1967.
68. Pickett, E. E. and S. R. Koirtyohann. *Spectrochim. Acta, 23B,* 235 (1968).
69. Dagnall, R. M., M. R. G. Taylor and T. S. West. *Spect. Lett., 1,* 397 (1968).
70. Dickinson, G. W. and V. A. Fassel. *Anal. Chem., 41,* 1021 (1969).
71. Skogerboe, R. K., A. T. Heybey and G. H. Morrison. *Anal. Chem., 38,* 1821 (1966).
72. Matousek, J. P. and V. Sychra. *Anal. Chem., 49,* 175 (1970).
73. Veillon, C., J. M. Mansfield, M. L. Parsons and J. D. Winefordner. *Anal. Chem., 38,* 204 (1966).
74. West, C. D. and D. N. Hume. *Anal. Chem., 36,* 412 (1964).
75. Hell, A. and S. Ricchio. 21st Anal. Chem. and Appl. Spectr. Conf. Cleveland, Ohio, 1970.
76. Aldous, K. M., R. M. Dagnall, B. L. Sharp and T. S. West. *Anal. Chim. Acta, 54,* 233 (1971).
77. Fleet, B., K. V. Liberty and T. S. West. *Anal. Chim. Acta, 45,* 205 (1969).
78. Ellis, D. W. and D. R. Demers. *Anal. Chem., 38,* 1943 (1966).
79. Manning, D. C. and P. Heneage. *Atomic Absorption Newsletter, 4,* 80 (1968).
80. Goto, H. K., K. Hirokawa and M. Suzuki. *Z. Anal. Chem., 225,* 130 (1967).
81. Murayama, S., H. Matsuno and M. Yamamoto. *Spectrochim. Acta, 23B,* 513 (1968).

82. Dagnall, R. M., G. F. Kirkbright, T. S. West and R. Wood. *Analyst, 95,* 425 (1970).
83. Warr, P. D. *Talanta, 17,* 543 (1970).
84. Ebdon, L., G. F. Kirkbright and T. S. West. *Anal. Chim. Acta, 58,* 39 (1972).
85. Dagnall, R. M., K. C. Thompson and T. S. West. *Anal. Chim. Acta, 36,* 269 (1966).
86. Kirkbright, G. F., M. Sargent and T. S. West. *Talanta, 16,* 245 (1969).
87. Kniseley, R. N., A. P. d'Silva and V. A. Fassel. *Anal. Chem., 38,* 910 (1963).
88. Cresser, M. S. and T. S. West. *Spectrochim. Acta, 25B,* 61 (1970).
89. Sychra, V. and J. Matousek. *Talanta, 17,* 363 (1970).
90. Sychra, V., P. J. Slevin, J. P. Matousek and F. Bek. *Anal. Chim. Acta, 52,* 259 (1970).
91. Smith, R. and A. E. Lawson. *Analyst, 96,* 631 (1971).
92. Fassel, V. A. and D. W. Golightly. *Anal. Chem. 39,* 466 (1967).
93. Browner, R. F., R. M. Dagnall and T. S. West. *Anal. Chim. Acta, 46,* 207 (1969).
94. Dean, J. A. and J. C. Simms. *Anal. Chem., 35,* 699 (1963).
95. Larkins, P. L. and J. B. Willis. *Spectrochim. Acta, 26B,* 491 (1971).
96. Robinson, J. W. *Spect. Lett., 2,* 37 (1969).
97. Herrmann, R. and C. T. J. Alkemade. *Chemical Analysis by Flame Photometry,* 2nd ed., John Wiley and Sons, New York, 1963.
98. Dean, J. A. *Flame Photometry,* McGraw-Hill, New York, 1960.
99. Mavrodineanu, R. and H. Boiteux. *Flame Spectroscopy,* John Wiley and Sons, New York, 1965.
100. Dean, J. A. and T. C. Rains. *Flame Emission and Atomic Absorption Spectrometry,* Vol. 1, "Theory," Marcel Dekker, New York, 1969.
101. Dean, J. A. and T. C. Rains. *Flame Emission and Atomic Absorption Spectrometry,* Vol. 2, "Components and Techniques," Marcel Dekker, New York, 1971.
102. Mavrodineanu, R. *U. S. National Bureau of Standards Miscellaneous Publication 281,* U. S. Gov't. Printing Office, Washington, D.C., 1967.
103. Mavrodineanu, R. in *Analytical Flame Spectroscopy: Selected Topics* (R. Mavrodineanu, ed.), MacMillan, London, pp. 651-753.
104. Stumm, W. and J. J. Morgan. *Aquatic Chemistry,* John Wiley and Sons, New York, 1970.

CHAPTER 4

THE SAMPLE

"The sample is usually only an infinitesimal part
of the total volume and is therefore representative
of the total mass only to the degree that uniformity
of chemical composition exists within the total mass.
In their natural state, surface and ground waters are
subjected to forces that promote mixing and homogeneity.
The fact that such tendencies exist, however, is not
sufficient cause for assuming that a body of water is
so well mixed that no attention to sampling technique
is required."

F. H. Rainwater and L. L. Thatcher[1]

SAMPLING PROGRAM

Regardless of the amount of care and attention
lavished upon the final instrumental analysis, these
data may be useless if the initial sampling program
is inherently faulty. Baldly stated, this fact is
reasonably obvious, yet very frequently this portion
of the analysis program is badly neglected, often--
unfortunately--because the sample collection and
chemical analysis steps are divorced from one
another. However carefully the analyst may analyze
and statistically circumscribe the water sample as
it arrives in the laboratory, the values obtained
will eventually be used to say something about the
parent water mass, not about the bottle of water.
In fact, far more errors are likely to be introduced
during the pre-analysis period and, whereas analysis
blanks and standards may delimit the analysis error,
pre-analysis, nonrandom errors may be almost impossible to detect. These problems are most clearly
appreciated where the analyst oversees both the

87

sample collection and the final data analysis.   In
view of the current concern over actual, potential
and (sometimes) supposed aqueous heavy-metal pollution, it should be the case that one person *is* in a
position to assess the significance of any given
batch of data.   Taking any metal in a species of
fish as a quite arbitrary example, the following
questions are clearly pertinent, but only the
sample collector may supply the answers:

1. Are the data biased by preferential collection
   in a particular area, season or at a particular
   time of day?
2. Were the samples taken from a particular organ
   of the fish; did one part (blood, for example)
   contaminate another; was the metal externally
   sorbed or physiologically bound?
3. Could subsequent handling, storage or other
   treatment have changed the *in situ* metal
   concentration and distribution?

The analyst (or the person who presents his data)
should also be aware of how one batch of numbers may
be knowledgeably fitted into the total dynamic system.
It may not necessarily matter how the values obtained
compare with some arbitrarily assigned "limits";
time and space variations are likely to be considerably more important.   Considerations such as these
argue for a sampling program carefully planned and
executed to answer specific logical environmental
questions.

Samples for Atomic Spectrometry

The quantity of discrete sample needed for subsequent AS analysis depends upon the concentration
range of the test element and the degree of sophis-
tication required of the analysis.   For conventional
flame analysis, some concentration/separation sequence will probably be mandatory.   At least 5 ml of
final solution per element are needed for reasonable
precision when standard nebulization is used.   The
initial sample volume required is predominantly a
function of the concentration system used, and this
can be estimated from first principles or by a
process of trial and error or, of course, by
following a published technique.
The nonflame atomization devices currently
available (considered in Chapter 7) require µl range
samples only, so that the preconcentration volume

required is correspondingly reduced. In our own work[2] we initially collect individual 10-ml samples (as illustrated in Figure 4.1) which are subsequently concentrated by a factor of 10. This provides sufficient material to determine five or six elements in replicate.

Figure 4.1.   Field collection and syringe filtration of water
samples for subsequent nonflame AA analysis.
Samples are taken directly from the samplers
illustrated in Figure 4.2.

There are no unusual restrictions imposed on aqueous samples for AS analysis, and small quantities (less than 2-3%) of solid particles in suspension may even be accommodated. The solids will not impede the normal nebulization and atomization of the liquid; whether or not the constituents of the solids are themselves atomized is an entirely different matter, and, because of the uncertainties involved this practice is not recommended. It is thus usual to prefilter or otherwise remove any coexisting solids. This is considered briefly below. Determination of the trace constituents of these solid phases by AS necessitates the addition of a dissolution or extraction step (see Chapter 5) and it is not possible to generalize about the quantity of sample needed. Analysis of nonaqueous liquid samples--petroleum residues, for example--is possible by standard AS techniques, although there are some rather severe requirements concerning standards.

Design of Sampling Program

It is hoped that some feeling for the importance of this topic has been conveyed by the previous discussions. It could be argued that the overall approach to steady state conditions increases from rivers through lakes to the marine environment. Particular deep oceanic water areas might well show little measurable change on a year-to-year basis, whereas freshwater streams could well require a continuous monitoring program. This is a crude generalization however, and even less applicable to reactive heavy-metal concentrations than to certain more conservative parameters. For example, pollutant metals may be concentrated and transported to oceanic areas by biota and will yield patterns totally alien to those generated by the passive transportation processes, although even in these cases such migration patterns might be predictable. In one sense, it is the varying and relatively greater quantities of contacting solid (inorganic and organic, see Reference 3) phases in freshwater environments which largely account for the gross deviations from steady state conditions in these waters. Using the concept of relative residence times (T_{rel}) given in Chapter 1, this index may be smaller or greater than unity where metals are removed from the system by (for example) sorption or, conversely, are mobilized by biotic uptake.

All in all it must never be forgotten that, unlike static rock or soil formations, all water systems are dynamic to varying degrees. Multiple sampling with subsequent combination to give a composite sample may be a statistically sound approach for some water bodies (but see Chapter 9); on the other hand this practice may totally mask some important localized perturbation. For all waters, the possibility of diurnal and seasonal cycles should also be considered. Estuarine flushing studies, for example, must take account of tidal mixing and annual freshwater discharge patterns.[4,5] O'Connor[6] has considered the general estuarine distribution of nonconservative constituents.

With regard to geographical coverage, the sampling program will be so completely dependent upon local conditions that nothing too useful can be included here. The majority of shallow streams and rivers will be vertically and horizontally homogeneous, and river sampling[7] is relatively straightforward. Conversely, the oceans and many lakes and estuaries are very likely to be stratified

to some degree. Construction of a reasonable depth-
sampling pattern in these cases may require considerable expertise and ancillary data (dissolved
oxygen or conductivity values, for example[8]). In
the case of specific localized discharge sites, a
logical pattern of samples should not be too difficult to evolve. It must be remembered that heavy
metals are often rapidly removed from the "active"
system by solids (see Reference 9 and Chapter 10),
so that soluble concentration values may decrease
very rapidly away from the source. This is one
example of the need to look at the phases coexisting
with the water; the sediment "sink" may not necessarily
always remain so.

The considerations noted so far represent something of an idealized approach to water sampling.
In practical terms the program will necessarily have
to take account of the availability of sampling
platforms, seasonal access and the whole gamut of
common logistical problems. This is inevitable;
however, some attempt should always be made to
understand and approach the ideal so that potential
limitations to the final data may be more clearly
appreciated.

Sample Collection

Practical methods for collecting water samples
have been well covered in numerous handbooks.[1]
Water is most usually collected as discrete samples
using sampler bottles of a variety of designs which
depend basically upon the volume of sample required,
the depth in the water column and various characteristics of the water body (current speed, for example).
This procedure is illustrated in Figure 4.2. The
greatest problem likely to be encountered here is
that of irregular contamination from metallic components (as discussed below), but other more subtle
errors may be introduced. As one example, the
physical sample collection process may distort the
water structure in some fashion. This same caution
applies equally to the passage of boats or to any
other nonstatic sampling platforms used.

Water samples may also be collected by direct
pumping, even from very deep water, although the
same potential contamination problems apply and may
even be intensified. Sampling in this manner affords
the analyst the ability to obtain a continuous record
with time and thus constitutes the basis of continuous monitoring systems. The need for such automated

Figure 4.2.   Withdrawing small (circa 10-ml) water samples from metal-free sampling bottles.

heavy-metal data collection in certain waters has
been well established and tentative on-line procedures are being developed for copper, mercury and
lead.[10] This type of real-time direct analysis is
ideal in nearly every respect and is in common use
for water parameters such as conductivity and dissolved oxygen contents. However, equivalent equipment for the detection of ions present in trace
quantities (within a variable matrix) is likely to
be exceedingly expensive into the foreseeable future.

Ice--sampled by a variety of coring devices--
and biota for subsequent trace-metal determinations
require subsampling to remove surfaces which have
been in contact with metal objects. Otherwise
collection methods for these phases are relatively
straightforward.

Ancillary Parameters

In most cases, soluble trace-metal data *per se*
are not going to convey very much useful information
about the quality of a specific body of water. Such
values will need to be fitted into the total environmental pattern so that perturbations and trends may
be recognized and analyzed. Once again it is not
possible to consider this topic here in detail because every case will be unique in some respect.
The following are common needs:

1. water movement properties: current velocities;
   mass transport data.
2. hydrographic properties: identification of
   water structures using conservative parameters;
   mixing, flushing rates.
3. Eh/pH data
4. suspended sediment characteristics: distribution; size ranges; mineralogy; organic content
5. factors affecting the chemical speciation of
   soluble metals: anion contents and concentrations; dissolved organic carbon distributions
6. factors affecting biota distributions:
   dissolved nutrients; productivity measurements;
   dissolved oxygen contents.

PRELIMINARY TREATMENT

Phase Separation

One major concern immediately following sample
recovery may be removal of coexisting particulates.
This topic bears some thought because the concentration of heavy metals bound to suspended sedimentary
material and biota (regardless of mechanisms) may
potentially exceed the soluble contents by several
orders of magnitude. Regardless of whether the
particulate fraction is also to be analyzed, it is
clearly important to decide whether phase separation
is to be performed immediately upon collection or
just prior to analysis. In the latter case the
equilibrium relationship of the *in situ* sample is
supposedly retained, although difficultly reversible
sedimentation of the solids, or biotic decompositions,
may occur during storage. Field removal of sediment
is a relatively easy operation so long as stringent
precautions against contaminations are taken.
Vacuum filtration apparatus is reasonably portable,
and there are a wide range of filter membranes
available in a variety of pore sizes. Cellulose or
glass-fiber filters are recommended for trace-metal
work. The analyst must be aware that filtering the
sample is potentially a serious source of random
error as it is in any operation where the sample is
brought into contact with large solid surface areas.
Table 4.1 shows per cent loss data[2] for metal tracers
passed through two commercial filtering rigs. A
Swinnex syringe filter in field use is illustrated
in Figure 4.1. Within individual limitations,
centrifugation may also be a viable field separation
technique.
Other early separation schemes will be required
under certain circumstances. Natural or introduced
hydrocarbon fractions can be separated from the
aqueous phase by solvent extraction in the usual
fashion. It should be noted by those environmental
chemists concerned with oil spills in natural waters
that some introduced petroleum products may also be
present in the water column as finely dispersed
particles. Ice sampling presents its own problems.
For example, sea ice contains discretely orientated
brine pockets which must be separated from the ice
sample without contamination of the latter.
Treatment of the sample in the field should
constitute but part of a well-planned scheme designed
to yield the maximum information about the physical

Table 4.1

Tracer Studies to Determine Losses During Filtration[2]

		Ag	Cd	Co	Pb	Zn
	% recovery[a]	99	98	–	70	–
Millepore Filter	% on filter	0	0		2	
	% on frit	0	0		18	
Swinnex Filter	% recovery[b]	95	100	90	84	92
	Required no rinses[c]	5	1	10	10	5
pH		8.0	7.8	7.8	7.8	7.8

[a] % recovery of tracer after passing 500-ml filtered seawater through standard rig with $0.45\ \mu$ membrane.

[b] % recovery after second rinse through filter.

[c] Approximate number of rinses required to prevent significant sorption losses.

and chemical distribution of the test element. One idealized separation plan is shown in Figure 4.3 (see also Chapter 5).

Storage and Contamination

In the absence of real-time analysis, water samples must inevitably be stored between collection and analysis, and great care must be exercised to limit undesirable changes which may occur during this period. First and foremost of these effects is contamination of the sample by the various con-tacting materials--the samplers, pumps, storage bottles, filter membranes, etc. This is a common problem, but one that is exceedingly difficult to combat. Several studies have been published showing total metallic impurities in commonly used containing materials. Some of the data determined by Robertson,[11] for example, are shown in Table 4.2. To some extent,

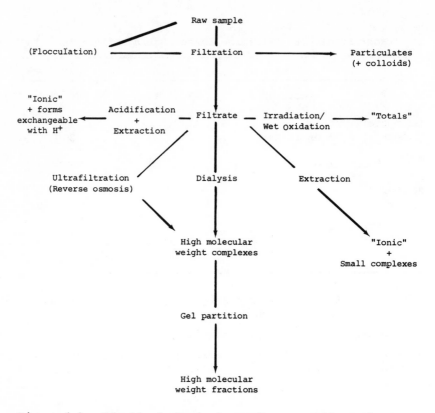

*Figure 4.3.   Idealized chemical species separation scheme
applicable to a natural water sample.*

data such as are given in Table 4.2 may be misleading
since, in most cases, the containing material does
not appreciably dissolve; rather the contaminants
are released from surface sorption sites into the
sample.   Table 4.3,[12] shows zinc contamination which
has been leached from various filter membrane
materials during filtration.   Data such as these
may aid in choosing equipment for pre-analysis handling, where choice is possible.   Needless to say,
very stringent cleaning of all containers and
apparatus is imperative.   Equally troublesome in
terms of irregular addition errors are impurities
in the chemicals used to effect the various pre-
analysis treatments considered in the following
chapter.   Some heavy-metal impurities found to be
present in common chelating agents, solvents and
other reagents are also included in Table 4.2.

Table 4.2

Concentrations (ppb) of Some Heavy Metals in Materials and
Reagents used for Containing and Treating Water Samples
(Data by neutron activation analysis from Reference 11)

	Ag	Co	Cr	Cu	Fe	Sb	Zn
a - Container Material							
Quartz Tubing[a]	< 0.1	1.1	602	0.03		38	33
Borosilicate Glass[b]	< 0.001	81			2.8×10^5	2900	730
Vycor Glass[b]						1.1×10^6	
Polyethylene	< 0.1	0.07	76	6.6	10400	0.18	28
b - Chelating Reagents							
Dithizone[c]	<10	1.2	<2000	420	<7000	0.8	1150
Cupferron[d]	3	0.68	< 200	160	< 600	0.3	7000
Thionalide[e]	4.6	5.1		100	< 300	3.7	120
APDC[c]	11	1.3		4	5000	1.9	1970
DEDTC[c]	<10	0.56			< 600	42	40
Oxine[c]	< 0.4	< 0.4	< 50		1000	6.0	< 100
c - Solvents							
Nitric Acid[e]	0.24	0.02	72	1.3	2	0.03	13
Hydrochloric Acid[e]	< 0.1	0.09	1.1	82	1	0.2	22
Ammonium Hydroxide[e]	< 0.1	0.01	< 0.04	6.0	< 0.1	< 0.01	2.3
CCl$_4$[fg]	< 0.005	0.003	< 50	0.12	9.8	0.33	1.2
CHCl$_3$[fg]	< 0.005	0.003	<100	0.29	1.6	0.05	2.1

[a] General Electric Co.
[b] Corning Glass Co.
[c] J. T. Baker Chem. Co.
[d] Eastman Co.
[e] Baker & Adamson, C.P. Reagent
[f] Mallinckrodt Co.
[g] Double distilled.

Table 4.3

Filter Membrane Contamination Data for Zinc
(μg/100-ml Aliquot Double Distilled Water)
(From Reference 12)

Treatment	Filter Material	Successive Filtrate Aliquots			
		1	2	3	4
a	Cellulose	0.06	0.01	0.01	0.00
	Glass	0.13	0.04	0.02	
	Silver	0.21	0.03	0.02	
b	Cellulose	0.00	0.00	0.00	0.00
	Glass	0.03	0.02	0.00	0.00
	Silver	1.13	0.06	0.20	0.01

[a]Untreated

[b]Treated with 10% HCl for 12 hrs and washed with double distilled water.

These values are not necessarily standard in any way, of course, but some contamination is to be expected from all containing materials and reagents. Repurification of chemicals is often a standard practice, but, unless the analyst is exceedingly skillful and well equipped, these precautions are likely to exacerbate rather than to improve the situation. Apparatus for the electrolytic purification of some reagents is now commercially available. Reagent blanks run through the entire analysis sequence are a constant requirement. The quantitative impact of such contaminations is, of course, almost totally dependent upon the type and quantity of reagent used. Thus ferric hydroxide coprecipitation concentration of trace heavy metals is not a highly recommended procedure,[13] but small acid additions may introduce analytically negligible impurities.

Adsorption of the analyte onto principally container surfaces produces a negative contamination[12] equally as serious as that originating from the additions discussed above. Such sorption effects have been particularly thoroughly documented in the case of silver (see Chapter 10) and potential problems with lead are illustrated in Table 4.1. These effects are quantitatively functions of the solution chemistry (pH, for example) but--of more practical importance--they are also functions of the nature of the solid surfaces. Glass and PVC bottles, for example, act differently and "aging" of containers may produce surfaces which increasingly promote adsorption of ions. Sorption appears to be universally retarded at low pH values, and acidification of samples to around pH 1-2 immediately on collection is becoming standard practice, although care must be exercised to ensure that no extraneous metallic impurities are added. Complexing reagents may also be added to hold the metals in solution. In this case possible effects upon the subsequent analytical performance must be clearly appreciated.

Acidification of water samples also inhibits precipitation of iron hydroxides from some waters with increase of pH on storage. Loss of carbon dioxide, for example, promotes this effect. Hydroxide precipitation must be avoided at all costs because of the potential removal of other heavy metals present in trace quantities by coprecipitation. Some people prefer to freeze samples for storage rather than use chemical additions. This practice has been criticized on the grounds that undesirable irreversible reactions result, but the evidence for this being of major concern is less than conclusive.

REFERENCES

1. Rainwater, F. H. and L. L. Thatcher. *Methods for Collection and Analysis of Water Samples,* U.S.G.S. Water Supply Paper 1454, U.S. Gov't. Printing Office, Washington, D.C., 1960.
2. Burrell, D. C., Valerie M. Williamson and M.-L. Lee. 4th Intern. Conf. Atom. Spectr., Toronto, 1973.
3. Eichholz, G. G., A. N. Galli, and L. W. Elston. *Water Res. R., 2,* 561 (1966).
4. Dyer, K. R. *Estuaries: A Physical Introduction,* John Wiley and Sons, London, 1973.
5. Bowden, K. F. in *Estuaries* (G. H. Lauff, ed.), Amer. Assoc. Adv. Sci., Washington, D.C., 1967.
6. O'Connor, D. J. *J. Sanit. Eng., 91,* SA1 (1965).
7. Klein, L. *River Pollution, I. Chemical Analysis,* Butterworths, London, 1959.
8. Welsh, P. S. *Limnological Methods,* McGraw-Hill, New York, 1948.
9. Preston, A., D. F. Jefferies, J. W. R. Dutton, B. R. Harvey, and A. K. Steele. *Environ. Poll., 3,* 69 (1972).

10. Ballinger, D. G. *Environ. Sci. Technol.*, *6*, 130 (1972).
11. Robertson, D. E. *Anal. Chem.*, *40*, 1067 (1968).
12. Burrell, D. C. Proc. 3rd Intern. Atom. Spectr. Cong.,
     Paris, 1972, pp. 409-428.
13. Burrell, D. C. *Anal. Chim. Acta*, *38*, 447 (1967).
14. Robertson, D. E. *Anal. Chim. Acta*, *42*, 533 (1968).

CHAPTER 5

PRE-ANALYSIS CHEMISTRY

"... we should stress the importance of not
being impressed by the elegance of technique.
Just as, in general, we are not interested in the
tool which a carpenter uses to drill a hole but
only in the result, so here. It is not particularly
important to catalogue methods; the relevance of
the result is the important factor."
                                    F. Morgan[1]

    It is very seldom that the concentrations of
trace heavy metals in natural waters can be determined by any analytical procedure without some
pre-analysis chemical treatment. Such a facility
would be, of course, highly desirable from the
standpoint of analytical speed and reduction of
contamination; both developmental work using the
various flameless and plasma atomization units
described in Chapter 7 and the *in situ* procedures
noted previously are tentative steps in this direction. However, implementation of such methods of
analysis could only yield one number per sample,
either a total datum or some species fraction, so
that one useful objective of pre-analysis manipulation
of the water sample is separation into constituent
phases and chemical groups such that the distribution
of the test element may be better understood. Basic
initial treatments for the removal of organic and
inorganic particulates, and organic soluble phases,
have been alluded to in the previous chapter; further
subdivision of the soluble fractions is considered
below. This type of physicochemical separation, if
attempted, may greatly aid in elucidating the behavior
of a particular element. Depending upon individual

101

circumstances, the analyst must decide whether
implementation of such a separation scheme is worthwhile, balancing the potential gain in information
against the possibility of introducing an unacceptable
degree of error. A more common requirement (and one
considerably less desirable) is the necessity for
pre-analysis treatment not primarily to add to the
sum of knowledge of the sample but as a mandatory
precursor to analysis by a chosen analytical method
having detection limits poorer than the natural
sample concentration range, or possibly to alleviate
some basic interference problem. The radiochemical
separations commonly required for neutron activation
analysis may be cited as one such example. In some
trace AS analysis, one equally routine requirement
is for the addition of excess quantities of a foreign
ion to act as a buffer to control emission interference
or to inhibit ionization of the analyte in absorption
analysis.

Flame AS analysis of heavy metals in natural
water samples, except under the most unusual circumstances, necessitates a pre-analysis concentration
step, frequently by a factor of up to x1000, and,
although this requirement has been considerably
alleviated through the use of nonflame atomization/
excitation procedures, it has seldom been completely
eliminated. These concentration steps usually
separate groups of elements, often by chemical
structure, although unfortunately this latter point
is frequently ignored or poorly understood. Regardless of potential fractionation of the sample in
this fashion, it cannot be over-stressed that any
or all of these processing steps may introduce grave
errors by, for example, promoting losses of the
analyte or through the introduction of random contamination. Some examples of the foreign heavy-metal
contents of common reagents have been given in Table
4.1. The final AS analysis can give only concentrations which exist at the time the sample reaches the
atomization cell, and the time and effort devoted by
the analyst to instrumental technique may be entirely
negated by various errors introduced during the
previous handling of the sample.

TREATMENT OF SOLID SAMPLES

The primary aim of this book is to consider the
treatment and analysis of liquid samples but, in
view of the control exercised on the soluble

concentrations of the heavy metals by the coexisting
biotic and inorganic particulates, it is necessary
to include, however inadequately, some consideration
of the solid phases. Both flame and nonflame AS
analysis at present require that sample be introduced in liquid form. A very few exceptions to this
rule are noted in Chapter 7, but these examples involve either specialized equipment seldom available
in the average analytical laboratory, or techniques
insufficiently proved for routine work. In general,
therefore, application of AS to the solid phases
necessitates an initial dissolution step which is
probably the greatest single source of contamination.
It is recommended that the analyst research the
possible use of other methods which are ostensibly
better suited to the direct analysis of solids (as
discussed in Chapter 2), particularly since the
concentration of heavy metals in the natural solid
phases (especially biota) may be one or more orders
of magnitude greater than in the contacting water.
However, there are equally good reasons for standard-
izing techniques, and preparation of an analyte
solution ensures sample homogeneity so long as the
specimen selected for dissolution is representative.

Dissolution

A great deal has been written about the dissolution of silicate material.[2] Prior to the widespread
acceptance of X-ray fluorescence analysis, all major
silicate elements were invariably determined
colorimetrically so that solution techniques are
well established and internationally standardized.
In recent years the classical methods have been
giving way to more efficient hydrofluoric acid
digestion in closed teflon-lined bombs and to rapid
borate fusions. Trace heavy-metal contents have
traditionally been obtained by arc spectrographic
methods (see Table 2.13), but AA analysis of the
prepared solutions is rapidly gaining ground.
Differential dissolution of some silicate material
is possible but not recommended.
Analysts chiefly concerned with pollutants in
natural water environments are likely to be far more
interested in the analysis of biota contents. Many
texts have discussed the various ways of treating
biological material in order to obtain the analyte
metals in a solution form suitable for AS analysis;
the monograph by Gorsuch[3] is recommended in particular.
Dry ashing, in which the sample is oxidized by heating

to a high temperature in air, is obviously not a
suitable treatment for the many metals of interest
which are volatilized at low temperature. Innumerable
variations of the oxidizing acid digestion procedure
are in common use, and, although the nitric-perchloric
acid mix probably predominates, it has been shown[4]
that other mixtures are about as efficient. One
important consideration for trace analysis concerns
the initial purity of the reagents (Table 4.2) and,
in this regard, nitric acid and hydrogen peroxide
are very suitable. It is seldom necessary to *completely* mineralize the biota sample. In our
laboratory, for example, we have used the nitric
acid vapor digestion of Thomas and Smythe[5] without
the addition of any further oxidant. This achieves
better than 90% destruction of plant material. One
other method which has gained wide popularity for,
in particular, the determination of traces of readily
volatilized trace metals in biological matter is a
low-temperature ash in oxygen within a high-frequency
field.[6,7]

One point which must be carefully considered is
the question of whether biota data are to be reported
on the basis of either initial wet or dry weight.
This question is of considerable pertinence in regard
to interlaboratory comparison of values. A wet-weight
basis has been traditional in many laboratories, but
this should probably be discouraged and replaced by
some standardized procedure such as freeze-drying.
This has been used, for example, for the mercury
data given in Table 10.6 (see also Reference 8).

Extraction

In some cases it may be possible to analyze the
fraction of interest in organic material by simple
extraction or leaching, rather than by application
of a total dissolution process. One example for
mercury is cited in Chapter 10. This is predominantly
a matter of convenience in preference to the more
complex treatments noted in the previous section,
although in some instances--such as removal of metals
weakly bound on, say, fish gill surfaces--isolation
of this fraction may add to the sum knowledge of the
sample. In this fashion a crude fractionation of
the metal into "sorbed" and "physiological" categories
is effected.

For inorganic sediment, it may be useful to
program various leaching reactions as a means of
determining the partition of metals between, for

example, reactive and lithogenous[9] or oxidizable and reducible[10] fractions. Total dissolution of silicate material will put into solution both the structural heavy-metal content and that held by various mechanisms on the particle surfaces. It is this latter fraction which is potentially mobile and hence of more immediate interest to the water chemist. There are many standardized methods for removing "exchangeable" alkali and alkaline earth metals from sediments based upon the techniques developed for soils, but uncritical extension of these to the heavy-metal suites is a questionable practice because of the obviously different modes of attachment and release discussed in part in Chapter 1. Thus treatment of marine sediment material with a concentrated transition metal salt will in general cause the release (exchange) of a greater fraction of a given heavy metal than will the conventional ammonium solution of the same strength.[11] No universal rules may be suggested here; the choice of treatment depends entirely upon the analyst's particular needs. However, it may be stated with some confidence that application of *some* extraction process to the sediment[12] will yield more information directly applicable to the problem of metal equilibria between these phases than may be obtained from a considerably more cumbersome total dissolution.

CHEMICAL SPECIATION OF
AQUEOUS SAMPLES

Very often the presence of several distinct chemical species of the metal in the water sample is tacitly ignored or, at best, an attempt may be made to obtain as close to a total analysis as possible. It was noted in Chapter 2 that for unequivocal "totals" a direct solid-sample method or a total consumption AS analysis is to be preferred. However, it may be unwise to ignore the individual chemical fractions since these will differ in "reactivity" and importance to the problem at hand. For example, the organic ligands may hold in solution heavy metals which would otherwise be removed to a sediment sink. A good illustration of this type of fractionation is provided by the data of Table 2.2 where water samples have been analyzed before and after acid treatment to release the complexed metal. The various preflame AS analysis chemical treatments––

chosen primarily to concentrate the metal fraction—may also act unwittingly to differentially isolate a particular chemical fraction. The analyst must be most vigilant in this respect, yet these methods may also be used in a controlled fashion to speciate the sample.

It is unlikely that the average reader will require a knowledge of the various soluble fractions beyond a basic subdivision into "simple ionic" and "organic" in addition to that associated with the coexisting particulate phases. Data of this type for marine zinc are included as an example in Table 5.1. It is theoretically possible, of course, to differentiate between the various inorganic ligand-metal complexes present in marine waters (Table 1.2) using various polarographic and specialized spectrometric procedures, but further consideration of this topic is beyond the scope of this book.

Oxidation and Reduction

It may be necessary to homogenize and break down complex species of the test metal to simple compounds more amenable to analysis by particular techniques. By these means, total metal contents in the original sample material are theoretically given. This type of treatment is required in most cases, for example, prior to direct anodic stripping voltammetry. For AS procedures, a pre-analysis oxidation should preface the various concentration steps since these are usually more applicable—or at least the course of the reactions is more predictable—if the analyte is in a homogeneous "ionic" form. Conventional oxidation reagents such as peroxydisulfuric acid[13] give perfectly acceptable results, but there is always the risk of contaminant additions. By far the best procedure is that of UV irradiation as developed by Armstrong *et al.*[14] This method has been quite successfully utilized in our laboratories for large volumes of water,[15] but in such cases the process is relatively slow. With the advent of flameless AA, only a μl-range sample volume is required so that quite small samples (5-10 ml) may be efficiently irradiated in batches.

Reduction treatments of samples are now commonly employed prior to the AA analysis of mercury and a range of pollutant metals which form volatile hydrides (see also Chapters 7 and 10). Freshly prepared $SnCl_2$ is usually used to reduce mercury compounds to the metallic vapor. Any of a number

Table 5.1

Recent Literature Data for Marine Zinc ($\mu g/\ell$)

Analysis	Treatment	Water Type	Extracted ("Ionic")	Non-dialysed ("Organic")	Particulate	Total	Reference
NAA	CSE	a	0.9-8.8	0.9-2.1	0.5-2.3	1.6-10.5	16
NAA	CSE	b	n.d.-7.3	n.d.-5.3	n.d.-2.0	n.d.-8.2	16
AAS	CSE	c	0.4-28.2		0.3-13.0		17
AAS		g				1.5-2.9	18
AAS	CR	a	0.7-10.0			1.0-11.0	19
ASV		c				30.0	20
AAS	CSE	f	1.1-6.2				2
AAS	CSE	f	10.9				21
ASV		d				0.3-1.2	22
ASV		e				0.3-3.4	22
AAS	CSE	h	3.0				23
AAS	CSE	i	3.0-5.0			3.0-5.0	24

NAA - neutron activation analysis
AAS - atomic absorption spectrometry
ASV - anodic stripping voltammetry

CSE - chelation/solvent extraction
CR - chelating resin

a Gulf of Mexico - surface water
b Gulf of Mexico - deep water
c Pacific Ocean - coastal water
d Pacific Ocean - surface open ocean water
e Pacific Ocean - deep water

f Atlantic Ocean - surface/coastal water
g Arctic Ocean - surface water
h Black Sea - surface water
i Pacific Ocean - estuarine water

of reducing agents (zinc, $NaBH_4$, etc.) are suitable
for the generation of the hydrides of, notably,
arsenic, selenium and antimony. Absorption analysis
via a gaseous phase serves as an efficient means of
separating these metals from any matrix material.
Concentrations of the metals are also easily
effected--several established methods for mercury
are noted in Chapter 10. Hydrides may be trapped
in, for example, liquid nitrogen and subsequently
released in a concentrated "burst." Christian[25]
has noted that a differential release of the hydrides
into the analysis tube is possible by this means.
Reduction treatment of sample material may also be
used as a separation technique regardless of the
subsequent atomization process. Several elements
of current concern (*e.g.*, selenium[26]) may be easily
reduced to the metal and separated from the matrix
by filtration.

Separation of Molecular Weight Fractions

The amount of dissolved organic matter in freshwater is about an order of magnitude greater than
that of marine waters[27] so that considerably more
is known about the associations of heavy metals with
the freshwater substances, particularly with the
humic acid fraction. In addition, the common initial
concentration processes applicable to fresh water
(*e.g.*, vacuum evaporation) are denied to marine
water samples because of the high matrix concentration of the latter. As one example, Barsdate[28] has
noted the use of both dialytic and electro-dialytic
concentration of metal-organic complexes from lake
waters, and also starch gel fractionation of several
broad-based, high molecular weight groups. Sephadex
column fractionation of the organic material in
natural fresh water is, in fact, quite commonly
employed (*e.g.*, Reference 29).
Evidence by several people[30,31] has confirmed
the association of heavy metals with the non-ionic
fraction separated and concentrated from seawater
by dialysis or direct solvent extraction. But there
has been very little work on molecular size-group
fractionation of this material. Ultra-filtration
(reverse osmosis membrane filtration) has proved to
be a highly useful concentration procedure for marine
soluble organics, although precipitation onto the
membrane surfaces may present problems. Figure 5.1
summarizes the results of one such fractionation
experiment on a seawater sample initially concentrated

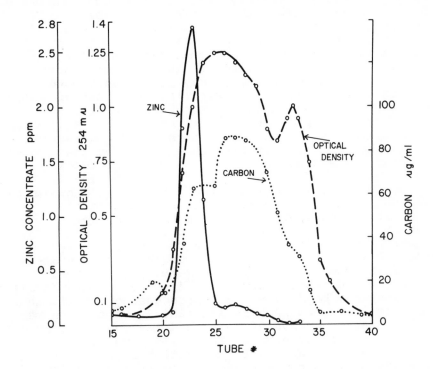

Figure 5.1. *Fractionation of organometallic complexes in sea-*
water. Zinc concentration, optical density (at
254 mμ) and organic carbon data for successive
4-ml fractions eluted from a G-50 Sephadex column.
The seawater sample was concentrated x200 prior
to fractionation. Data from Reference 32.

by reverse osmosis from 4000 to 20 ml and shows the
association of zinc with successive 4-ml fractions
eluted from a Sephadex G-50 column.[32] The metal
appears here to be associated with the molecular
weight fraction in excess of 10,000.

SEPARATION AND CONCENTRATION
OF SOLUBLE METAL SPECIES

Even with the most complete optimization of all
instrumental components and techniques, the ranges
of most heavy-metal constituents of natural waters

will usually lie outside the AS analytical working
range. This is certainly true for most continuous
flame analysis and, although discrete flameless
analysis is improving this situation, the full
potential of these techniques lies substantially
in the future so that a pre-analysis concentration
step is frequently mandatory. In addition, particularly for emission work, it may be impossible to
separate the analyte analysis line from that of some
coexisting concomitant, and the concentration step
may have to incorporate (or be replaced by) a closely
controlled separation. All of these procedures involve the risk of potential sample loss or the addition of contamination--as with all the processing
steps considered in this chapter. It is clearly
impossible here to discuss the pitfalls of individual
procedures but, in general, the best approach is
that which accomplishes the job using the least
number of steps and reagents. Unlike the radiochemical manipulations used in conjunction with
neutron activation, it is not possible to accurately
determine the extent of (for example) loss of the
test metal directly, but the efficiency of standardized
routines may be determined by processing a suitable
radiotracer where such is available. Of the innumerable general concentrating procedures, only a very
few are regularly used to preface AS analysis.
Techniques such as electrodeposition or freeze-drying,
for example, do not put the sample in a convenient
form for subsequent AS nebulization. Direct evaporation removal of water may be usefully applied to
freshwater, and large concentration factors (*e.g.*,
Reference 33) may be routinely attained.

One of the oldest separation/concentration methods
is that of precipitation, including co-precipitation
and crystallization. These routines may be made
specific for certain elements or groups and are in
widespread use. But the commonest methods chosen
to accompany AS analysis utilize a controlled partition of the metal *ions* between two phases. Thus, at
equilibrium for a species of R distributed between
phases a and b:

$$\mu_a = \mu_b \qquad (5.1)$$

and since the chemical potential (μ) of R in phase a
takes the form:

$$\mu_a = \mu_a^o + RT \ln (R)_a \qquad (5.2)$$

(where μ_a^o is a constant at a specified temperature and pressure) the activity ratio:

$$\frac{(R)_a}{(R)_b} = \exp \frac{\mu_a^o - \mu_b^o}{RT} = \text{Const.} \qquad (5.3)$$

Solvent Extraction

Liquid-liquid solvent extraction constitutes the single most useful separation method for all chemical analysis, and many excellent texts and reviews are available (*e.g.*, References 34-39). Major efforts in this field to date have been devoted to the development of specific reactions, for use in colorimetry, for example. It may be found necessary in some cases to separate individual elements for flame emission analysis also, but the requirements for AS analysis in general are somewhat different; most frequently broad group separations suffice and very often the prime requirement is for a *concentration* of the analyte.

For the distribution of the analyte species (R) between two immiscible liquids a and b, the partition coefficient (P) is given by:

$$P = \frac{(R)_a}{(R)_b} \qquad (5.4)$$

This is a true thermodynamic partition. In practice, the Nernst partition isotherm is defined:

$$D = \frac{[R]_a}{[R]_b} \qquad (5.5)$$

in concentration terms, which holds approximately true for most practical purposes. Here D is an "extraction coefficient" or a "distribution ratio" as distinct from the ideal partition coefficient P.

To produce a favorable extraction of the metal ion (R^{n+}) of interest from one phase (aqueous) to the second (organic) phase, it is usual to complex the metal with an organic ligand (L^-), commonly by a chelate reaction. This is the principle of most of the colorimetric reactions given in Table 2.7. In such cases, Equation 5.5 may be written:

$$D = \frac{[RL_n]_{org}}{[R^{n+}]_{aq}} \qquad (5.6)$$

This also represents an ideal situation, however, since most chelating agents function as weak acids and dissociate to some extent in the aqueous phase, so that the actual distribution of R^{n+} depends upon equilibria of the type:

$$HL_{aq} \rightleftharpoons HL_{org} \qquad (5.7)$$

$$HL \rightleftharpoons H^+ + L^- \qquad (5.8)$$

$$R^{n+} + nL^- \rightleftharpoons RL_n \qquad (5.9)$$

$$RL_{n \cdot aq} \rightleftharpoons RL_{n \cdot org} \qquad (5.10)$$

The final distribution is a function of the nature and concentration of the specific chelating reagent, the character of the organic solvent, the *rate* of chelate formation and--most importantly--the pH of the extraction (see Equation 5.8). Figure 5.2 shows the relationship between length of "contact time" and the extraction efficiency from seawater for the system DADDC-CCl$_4$ (see Appendix 2 for full chemical names of common abbreviations). In this example a slow, mechanical rotary mix has been employed. The importance of the pH upon the solvent extraction efficiency is generally well known. Figure 5.3 is one arbitrary example which illustrates TFA-toluene extraction of iron and indium from seawater[40] as a function of pH, and the optimum pH ranges for the extraction of various primary dithizonates are given in Figure 5.4. Such pH dependence may be utilized for the selective extraction of particular metals or groups of metals (common practice for colorimetry, see Table 2.7), but apart from certain AE techniques, it is usual to attempt to extract broad spectra of metals for subsequent analysis by AS so that determination of an optimum extraction pH is a primary requirement. It is also of importance to note that buffering chemicals must not be allowed to contaminate the sample and that the "equilibrium pH" may be shifted to a marked degree by the acidity of the

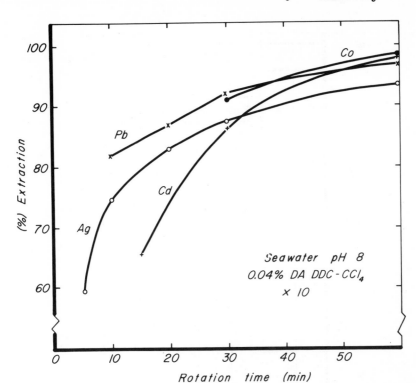

Figure 5.2. Extraction efficiency as a function of the mixing time (slow, vertical mechanical rotation) for the system DADDC-carbon tetrachloride from seawater at pH 8.

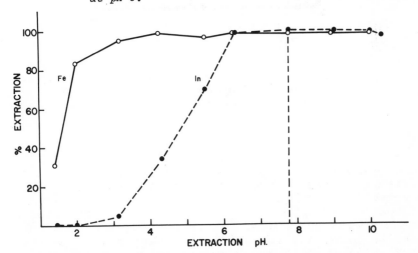

Figure 5.3. Effect of pH on solvent extraction. pH vs. % extraction of iron and indium from seawater using the system TFA-toluene. See Reference 40.

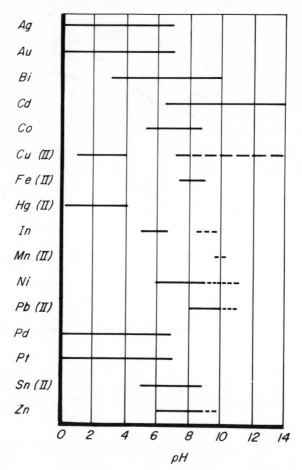

Figure 5.4. pH ranges for the quantitative extraction of the
primary dithizonates (broken line for copper is
for the secondary complex) in carbon tetrachloride
or chloroform (dotted line). Data primarily from
Reference 37.

chelating agent. Most chelation-solvent extraction
systems in common use for separating and concentrating
suites of trace heavy metals from natural waters
function best over low pH ranges. This conveniently
meshes with the acidification storage treatment of
the samples, and is useful if an approximation to
the total soluble metal content of the sample is
required. There is, however, some advantage to
extracting at the *in situ* pH (*e.g.*, Reference 24)

in order to separate the "ionic" fraction from organometallic species which may be present.

The extraction efficiency (E) for removing a metal complex from the aqueous to the organic phase is given by:

$$\%E = \frac{100\ D}{D + V_{aq}/V_{org}} \qquad (5.11)$$

where V_{aq} and V_{org} are the volumes of the phases contacted. It is possible to increase the total amount of complex extracted from the aqueous phase by performing repeated extractions with fresh organic material. This may be accomplished batchwise or by continuous methods,[41-43] but such procedures are seldom used for routine metal extraction. In most cases E may be calculated with little trouble, or standards may be processed in the same fashion as the samples to compensate for less-than-total extractions. The least preferable route (but, unfortunately, one commonly followed) is to assume that the extraction efficiency approximates 100%. Other common practical problems concern adsorption of the metal species onto the walls of the separatory containers, emulsion formation at the interface and--a common difficulty with seawater--precipitation due to an induced pH shift. Although batch separation will be found more convenient for most analytical purposes, it *is* possible to automate the mixing step in a relatively simple fashion (for example, for ship-board operation[44]). Where lighter-than-water solvents are chosen, it is convenient for subsequent flame AS analysis to separate and aspirate directly from narrow-necked flasks. Considering further the role of solvent extraction *concentration*--as against the more usual separation function--it should be noted from Equation 5.11 that the concentration factor increases with D, but decreases as the volume ratio (V_{aq}/V_{org}) is decreased. A practical limit to the latter might well be dictated by the solubility of the organic solvent in the aqueous phase, however, since the above simplified theoretical discussion has assumed a perfect immiscibility seldom approached in practice.

The utility of selecting a chelating agent suitable for sequestering a broad spectrum of metals has been emphasized above. Figure 5.5 illustrates the range of metals amenable to chelation by five of the more commonly employed nondiscriminating reagents

Oxine

Ti	V		Mn	Fe	Co	Ni	Cu	Zn	Ga			
Zr		Mo				Pd		Cd	In	Sn	Sb	
Hf								Hg	Tl	Pb	Bi	A

Cupferron

Ti	V			Fe	Co		Cu		Ga			
Zr	Nb	Mo							In	Sn	Sb	
	Ta	W								Pb	Bi	B

Dithizone

		Cr[a]	Mn	Fe	Co	Ni	Cu	Zn				
						Pd	Ag	Cd	In	Sn		
		W[b]	Re[b]			Pt	Au	Hg	Tl	Pb	Bi	C

DEDTC

	V	Cr	Mn[c]	Fe	Co	Ni	Cu	Zn	Ga		As	
		Mo			Rh	Pd	Ag	Cd	In	Sn	Sb	
						Pt	Ay	Hg	Tl	Pb	Bi	D

APDC

	V		Mn[c]	Fe	Co	Ni	Cu	Zn			As	
								Cd		Sn	Sb	
								Hg		Pb	Bi	E

Figure 5.5.   *Solvent extraction separation/concentration reactions: metals quantitatively complexed by the more common nondiscriminatory chelating reagents (refer to Appendix 2).*

A. Oxine. Data principally from References 34, 37, 45; refer to these sources for applicability to specific oxidation states.

B. Cupferron. Data principally from References 34, 37.

C. Dithizone. Data principally from Reference 46. See Figure 5.4 for extraction pH range of primary dithizonates.

D. DEDTC. Data sources as for E.

E. APDC. Documented applications of this reagent in conjunction with AS analysis are legion; this is a compendium of recent data.

[a] *See Reference 47.*
[b] *See Reference 45.*
[c] *See Reference 48.*

(for problems associated with the extraction of manganese see, for example, Reference 48), and the pH ranges over which a quantitative extraction of a number of metals may be effected using D DDC (in carbon tetrachloride) are listed in Table 5.2. A

Table 5.2

pH Range (1-12) for 100% Extraction of Metals by DADDC in CCl₄
(1:1 ratio; 5 min. shaking period). Data from Reference 49.

Element	pH Range	Element	pH Range
Ag (I)	1-12	Mo (VI)	1-4.5
As (III)	1-5.5	Ni (II)	2-12
Bi (III)	1-12	Pb (II)	1-12
Cd (II)	1-12	Pd (II)	1-12
Co (II)	3-12	Pt (II)	1-12
Cu (II)	1-12	Sb (III)	1-9.5
Fe (II)	3-10	Se (IV)	1-5.5
(III)	3-10	Te (IV)	1-8.5
Ga (III)	4.5-6.5	Tl (I)	3.5-12
Hg (II)	1-12	(III)	1-12
In (III)	1-9.5	V (V)	4-5.5
Mn (II)	6.9	Zn (II)	2.5-12

review of nonspecific systems especially applicable to seawater has also been given by Joyner *et al.*[50] There are, of course, a multitude of more exotic chelating ligands which may be preferable for certain applications and particularly where more discriminatory reactions are sought; Table 5.3 gives an abbreviated selection (see also the data in Reference 51 on selenazone). It must be remembered that many of these otherwise exemplary complexing organics are unstable to a lesser or greater extent and must be freshly prepared at frequent, regular intervals. And if commercial reagents are used, the analyst should be on guard for heavy-metal impurities (Table 4.1); oxine seems to be a perennial offender in this respect.

The best solvents might appear *a priori* to be those which yield the most favorable extraction distribution ratio (Equation 5.6). However, for AS flame atomization the practical choice is most often dictated by the relevant combustion properties. This is the basic reason for the popularity of MIBK,

Table 5.3

Elements Complexed by Some Less Commonly Used
Nondiscriminatory Chelating Reagents
(see Appendix 2) under Ideal Conditions
Data from References 34, 37 except as otherwise cited

Element	Dithiol[a]	PAQA[b]	HAA[c]	S-oxime[d]	IOTG[e]
Ag				X	X
Au					X
Bi	X			X	X
Cd		X			
Co[f]			X[g]		
Cu		X		X	X
Fe[f]		X	X	X	
Hg					X
In[f]			X		
Mn[f]			X[g]		
Mo[f]	X			X	
Ni		X		X	
Pb				X	
Pd		X		X	
Re[f]	X				
Sn[f]	X				
V[f]				X	X
W	X				
Zn		X			

[a] Acidic media.
[b] Data from Reference 52.
[c] Frequently requires no additional solvent.
[d] Data predominantly from Reference 53.
[e] Acidic media; requires no additional solvent; data from Reference 54.
[f] See source references for application to specific oxidation states.
[g] With oxidant.

since the extraction properties of this solvent
are not especially impressive. Published applications
of the system APDC-MIBK are legion (*e.g.*, References
55 for fresh waters and References 17 and 56 for
marine waters) and studies of the extraction of other
chelate complexes into MIBK are very common also
(*e.g.*, oxine,[57] dimethylglyoxime[58]). Takeuchi *et
al.*[59] have given a general discussion on the extraction ranges of various chelates with MIBK. Several

other ketones and a few esters have also been extensively used as solvents for pre-AS analysis
extractions; ethyl propionate and acetyl acetone
are prime examples.[60],[61]

However, choosing the organic phase solely, or
even predominantly, on the basis of the exhibited
combustion properties is not always sound analytical
practice; the system selected must be efficient and,
more importantly, readily reproducible. In many
cases it may be advantageous to use a superior
organic such as carbon tetrachloride or chloroform
and to back-extract into a weak acid solution, even
though this adds an additional processing step and
negates the signal enhancement effects realized
through nebulization of the organic solution (see
Chapter 7). In fact, the role of the organic
solvent is seldom closely considered by many
analysts. This is unfortunate, and the reader
is urged to refer to theoretical treatments such
as that by Wakahagashi *et al.*[62] on the extraction
of acetylacetonates by various solvents. Alimarin
and Zolotov[63] have subdivided chelate reactions on
the basis of the valence/coordination number ratio
of the central atom into coordination-saturated and
-unsaturated compounds. The latter complexes coordinate water molecules and are often poorly extracted
unless concomitant displacement of this water occurs,
i.e., by suitable choice of a (usually) oxygen-
containing solvent. As one example: for the
efficient extraction of TFA chelates of zinc and
cobalt, it was found necessary to add additional
isobutylamine.[40]

The analytical signal enhancement of organic
over aqueous solutions in flame AS has been referred
to above and will be considered in more detail in
Chapter 7. This phenomenon might be termed, in
part, a positive transportation interference
(Chapter 8) since it is partially due to a decrease
in solution viscosity. One other potential analysis
interference effect associated with pre-analysis
solvent extraction which should be looked for is
attributable to differential atomization of the
metal from the various chelate complexes. This is
a typical condensed-phase interference and has been
documented for mercury analysis by Law.[64] A comprehensive review of the application of chelate complexes
to flame AS technique has been given by Robinson
et al.[66]

The introduction of flameless AS analysis has
changed the extraction/concentration requirements
for natural water samples to some extent. First

and foremost, the primary need for a concentration
rather than a separation has been diminished.
Secondly, since the organic phase is no longer
supporting flame combustion, the complexing-extraction
system may be selected on other, more profitable
grounds. The initial furnace designs were not well
suited for atomizing organic solutions directly, but
this has been remedied with some second generation
designs as noted in Chapter 7. Carbon filaments
are particularly useful for organic phase samples.[24]
Since all of the present nonflame cells accept μl-
range samples, pre-analysis chemistry may be performed
on very small volumes, and extraction in tubes with
centrifugation separation is one recommended procedure.

Ion Exchange

     This separation involves partition of the analyte
between a solid and a liquid (aqueous) phase. The
role of sorption and ion exchange mechanisms in removing metal species from solution has been considered
previously, and it has been noted that this phenomenon
in nature comprises many individual physicochemical
reactions, including the probability of direct precipitation onto the substrate. However, for analytical
applications, it is necessary to "idealize" the reactions (or at least make the processes reproducible)
by using standardized solid phases of known behavior.[66]
Most usually, synthetic resins are used for the separations, and hence the common applications are termed
"resin ion-exchange." By way of example, the exchange
reaction of the R^{n+} metal ion on a sulfonic acid-type
resin may be considered:

$$R^{n+}_{aq} + nH^{+}_{resin} \rightleftarrows R^{n+}_{resin} + H^{+}_{aq} \qquad (5.12)$$

so that the selectivity coefficient K_s is given by:

$$K_s = \frac{R^{n+}_{resin} \, H^{+n}_{aq}}{R^{n+}_{aq} \, H^{+n}_{resin}} \qquad (5.13)$$

For any given resin material, variations in K_s for
individual metals will lead to a series analogous
to the exchangeable cation series of clay mineral
or soil classification terminology. A more practical
coefficient (K_D) may also be found helpful for

characterizing the partition of a given metal ion between solution and resin:

$$K_D = \frac{[R^{n+}]_{resin}}{[R^{n+}]_{aq}} \tag{5.14}$$

This is equivalent to the extraction coefficient for liquid-liquid extraction defined by Equation 5.5.

Everyone will be familiar with the routine application of resin exchange to purify or soften water, and most analytical procedures are conducted similarly using the resin packed in columns, except that continuous flow is not conducive to quantitative separation and discrete samples are usually run in a batch-wise fashion. The sample is added to the column, washed and eluted using a solvent best suited to both the resin and the metal (see, for example, Table 5.4 and standard texts[66,67]). If a concentration is desired, the volume of eluent must be smaller than the original sample volume. In addition to the relative extraction coefficient values, these separation operations are a function of the rate of the exchange process (which depends upon diffusion both within the resin material and across the liquid-solid boundary) and the resin exchange capacity. The latter parameter is again analogous to the well known cation exchange characteristic of clays, but resins utilized for analytical separations have very large capacities, and this term is not usually limiting.

Simple cation and anion exchange resins are usually used to separate broad chemical groups, or for applications such as the removal of particular interfering anions. For heavy-metal separations--and particularly if a concentration is required as is usually the case for pre-AS analysis separations--it is frequently advantageous to employ a chelating resin, that is, a resin to which has been coupled a suitably active organic ligand.[68] This type of resin has been commercially available and widely used for a range of waters (*e.g.*, industrial wastes,[69] sea-water[70,71]) for many years (Table 5.4). There is, however, a great deal to be said in favor of the analyst preparing his own chelating resins by coupling the organic group to some suitable base material.[72] In this way a complex may be tailored to a specific application and potentially troublesome contaminants may be better controlled. Figure 5.6 shows AA recorder traces for zinc-spiked seawater (A)

Table 5.4

Some Resin-Exchange Extractions Suitable for Concentrating Heavy Metals from Natural Waters

Element	Resin	Adsorption pH	Eluent	Recovery (%)	Reference
Ag	1	7.6	2N HNO$_3$	90[b]	70
	2	1.0	Acetone - HNO$_3$	100	73
Bi	1	9.0	2N HClO$_4$	100[b]	70
Cd	1	7.6	2N HNO$_3$	100	70
	4	7.6	0.1 KCN	91	74
Co	1	7.6	2N HCl	100	70
	3	10.3	6N HCl	75	75
Cr (III)	1	5.0	2N HNO$_3$	10[b]	70
Cu	1	7.6	2N HNO$_3$	100[b]	70
	3	10.3	6N HCl	100	75
	4	7.6	EDTA	100[b]	74
Fe	1	5.2	1N HNO$_3$	100	69
In	1	9.0	2N HNO$_3$	100[b]	70
Mn	1	9.0	2N HNO$_3$	100[b]	70
Mo[a]	1	5.0	4N NH$_4$OH	100[b]	70
Ni	1	7.6	2N HNO$_3$	100[b]	70
	3	6.0	6N HCl	100	75
Pb	1	7.6	2N HNO$_3$	100[b]	70
	1	5.2	1N HNO$_3$	100	68
	4	7.6	2N NH$_4$Ac	40[b]	74
Re[a]	1	7.6	4N NH$_4$OH	90[b]	70
Sn (IV)	1	7.6	2N HNO$_3$	60[b]	70
Tl (I)	1	7.6	2N HNO$_3$	50[b]	70
V[a]	1	6.0	4N NH$_4$OH	100[b]	70
W[a]	1	6.0	4N NH$_4$OH	100[b]	70
Zn	1	7.6	2N HNO$_3$	100[b]	70
	1	5.2	1N HNO$_3$	100	69
	4	7.6	2N NH$_4$Ac	65[b]	74

1 - Chelex-100 (purified form) or Dowex A-1 chelating resin
2 - anion exchange resin
3 - polyamine (TEPA)-polyurea resin
4 - Chitosan

[a] As anion
[b] From seawater

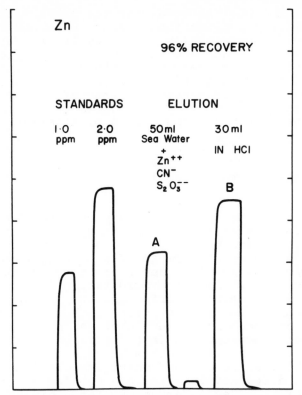

Figure 5.6. *Chelating-resin column extraction of zinc from a*
seawater matrix. AA signal tracers for zinc
standards, seawater plus zinc spike (A) and zinc
eluted from dithizone resin (B). Concentration
factor x1.7. After Reference 15.

and after resin extraction (B). The concentration
factor in this example is 5/3 and the resin comprises
dithizone coupled to a modified carboxyl cellulose
matrix, following the procedure advocated by Bauman
et al.[76] It must be admitted, however, that these
columns are both time consuming to prepare and
difficult to regenerate.

If the primary use of the resin column is to
effect a concentration rather than a separation,
additional practical problems emerge, as was the
case also with the solvent extraction procedures
described above. For example--and assuming that the
exchange capacity of the column is not exceeded--
small quantities of resin in relation to the original

sample volume must be used, and the extraction is
very slow. In applications where the column recovery
efficiency is potentially limiting, it is possible
in some cases to ash the resin bed and analyze the
latter; but this represents something of a retrograde
procedure for subsequent AS analysis. One other
technique worthy of mention is that of precipitation
ion-exchange[77],[78] whereby the *matrix* is precipitated
within the column and the metals of interest eluted
in a following step.

Precipitation-Crystallization
Reactions

     Separation methods based on precipitation removal
of the elements of interest might appear *a priori*
to be obvious and useful procedures. However--for
various reasons--as a pre-analytical preparation,
these reactions are less desirable than the partitions
discussed above. Direct precipitation is seldom used
for elements present in trace amounts; the principal
problem being one of adsorption or chemical incorporation of concomitant ions with the primary
precipitate complex. In fact, it is this latter
process of coprecipitation--embracing a variety of
physicochemical reactions--which is most often used
per se as discussed in principle by, for example,
Minczewski.[79]   (See Reference 80 for a discussion of
this phenomenon as it relates to one specific metal
and substrate.) The most commonly utilized anion
groups, together with elements reportedly thus
quantitatively coprecipitated, are given in Table 5.5.
Since the end-product of these reactions is a solid
phase, these techniques are best suited to analysis
schemes not requiring liquid samples. The precipitates may, of course, be redissolved and in a
considerably smaller volume to effect a concentration,
but each additional processing step tends to detract
from the final accuracy of the data. Precipitation
in situ of hydrated ferric oxide has frequently been
used to concentrate trace transition metals from
seawater.[56]   Such scavengers are usually presumed
to act as charged colloids, but the processes are
far from simple and have been considered in some
detail by Kim and Zeitlin.[81]   Manganese (IV) oxide--
produced, for example, by *in situ* permanganate
oxidation of an acidic manganese (II) solution--has
also found wide application (see Reference 82 and
Table 5.5). The practical problems associated with
all such "classic" coprecipitations are predominantly

Table 5.5

Some Commonly Utilized Coprecipitation Complexes
Data predominantly from Reference 79

Element	(Hydr)oxide	Sulfide	Carbonate[a]
Ag	X	X	
As		X	
Au	X		
Bi	X[b]	X	
Cd	X	X	X
Co	X	X	X
Cr	X		
Cu	X	X	X
Fe	X[b]	X	X
Ga	X[b]		
Hg	X	X	
In	X[b]	X	
Mn	X	X	
Mo	X[b]	X	
Ni	X	X	
Pb	X[b]	X	X
Pd	X	X	
Pt	X		
Rh	X	X	
Ru	X		
Sb	X[b]	X	
Ti	X		
Tl	X	X	
V	X		
Zn	X	X	X

[a] Na_2CO_3 added to seawater medium. Data of Reference 83 only; this tabulation is not exclusive.

[b] MnO_2 coprecipitation; data of References 82, 84 only.

two-fold, namely, the necessity for further treatment of the end-product (alluded to above) and the inevitable addition of foreign material with the large amount of carrier needed. Obviously this latter problem is likely to be particularly severe if one transition metal is employed to separate another.

Organic reagents are frequently used in place of the inorganics considered above. In these cases, chelating reagents are commonly employed so that the metal is complexed in a like fashion to that described for solvent extraction. Over certain pH ranges these organometallic compounds are insoluble in aqueous solution and precipitate. A selection of the more commonly used nondiscriminatory reagents is given as Figure 5.7 with the heavy metals reported to be "quantitatively" extracted.   It is not, of course,

Tannic Acid

Ti		Cr	Mn	Fe	Co	Ni	Cu	Zn	Ga		As
Zr	Nb							Cd			
	Ta							Hg		Pb	

A

Oxine

Ti	V	Cr	Mn	Fe	Co	Ni	Cu	Zn	Ga		
Zr		Mo				Pd		Cd	In		Sb
		W								Pb	Bi

B

Thionalide

							Cu				As
						Pd	Ag	Cd		Sn	Sb
						Pt	Au	Hg	Tl	Pb	Bi

C

S-Oxime

	V		Mn	Fe	Co	Ni	Cu	Zn			
						Pd	Ag	Cd			
								Hg		Pb	Bi

D

Cupferron

Ti	V			Fe					Ga		
Zr	Nb	Mo								Sn	
	Ta	W									Bi

E

Figure 5.7.   Reagents suitable for the precipitation separation of the heavy metals shown (under ideal circumstances).   Data from Reference 45.
A. Tannic acid.   Acidic media.
B. Oxine.   See Reference 45 for precipitation pH range and References 85, 86 for application of mixed oxine, thionalide and tannic acid reagent.
C. Thionalide.   Acidic media.   See Reference 87 for co-crystallization utilization.
D. S-oxime.
E. Cupferron.   Acidic media.

a practical proposition to quantitatively handle
only the precipitates of the metals present in trace
amounts, so that again carriers are generally a
necessity. This type of processing has found particular favor for subsequent spectrographic analysis
where not only are solid samples advantageous, but
internal standardization additives are commonly
required and the carrier may serve a dual purpose.
The mixed oxine-tannic acid-thionalide reagent of
Mitchell[88] has proved very popular and one example[86]
of the use of this mix is included in Table 2.6. In
this case an In/Pd carrier functioned also as the
internal standard. The precipitation reactions of
thioacetamide closely parallel those of the sulfides,
as summarized in Table 5.6[89] for both acidic and
ammoniacal solutions. Here Sn/In is the recommended
carrier if a final spectrographic analysis is con-
templated. A very common use of these organic
reagents has been for the concentration/separation
of trace metals from seawater, using the 5,7-dibromo
derivative of oxine, for example.[90] Tappinger and

Table 5.6

*Precipitation Concentration of Aqueous Heavy Metals
in the μg/ℓ Range Using Thioacetamide
Data from Reference 89*

	Element	Concentration (μg/ℓ)	Recovery (%)
Acid precipitation	As	>10	100
	Cd	> 5	100
	Cu	> 5	100
	Mo	>10	< 80
	Pb	> 5	100
	Sb	> 5	100
Ammoniacal precipitation	Ag		v.i.
	Co		v.i.
	Cr	> 0.5	100
	Fe	> 5	> 90
	Mn		v.i.
	Ni		v.i.
	Ti	>10	> 90
	Zn	> 5	> 95
	Zr	> 0.5	100

v.i. - variable/incomplete recovery.

Pickett[91] have discussed the application of several
organic coprecipitants over the period 1960-65 (see
also Reference 92). Weiss and co-workers[93],[94] have
published several illustrative examples of the
closely similar technique of co-crystallization in
which the organic reagents are dissolved in a water-
miscible, volatile organic solvent. The latter
subsequently evaporates to yield precipitation
(crystallization) of the reagent with concomitant
co-crystallization of the analyte metals.

REFERENCES

1. Morgan, F. *Medical Science and the Law, 6,* 155 (1957).
2. Dolezal, J., P. Povondra, and Z. Sulcek. *Decomposition
    Techniques in Inorganic Analysis* (English Translation
    from Czech, D. O. Hughes, P. A. Floyd, and M. S. Barratt,
    eds.), Iliffe, London, 1968).
3. Gorsuch, T. T. *The Destruction of Organic Matter,*
    Pergamon Press, Oxford, 1970.
4. Knauer, G. A. *Analyst, 95,* 476 (1970).
5. Thomas, A. D. and L. E. Smythe. *Talanta, 20,* 469 (1973).
6. Gleit, C. E. and W. D. Holland. *Anal. Chem., 34,* 1454
    (1962).
7. Gleit, C. E. *Anal. Chem., 37,* 314 (1965).
8. LaFleur, P. D. *Anal. Chem., 45,* 1534 (1973).
9. Chester, R. and M. J. Hughes. *Chem. Geol., 3,* 199 (1967).
10. Presley, B. J., Y. Kolodny, A. Nissenbaum and I. R.
    Kaplan. *Geochim. Cosmochim. Acta, 36,* 1073 (1972).
11. Burrell, D. C. "IAEA Symposium on the Interaction of
    Radioactive Contaminants with the Constituents of the
    Marine Environment," Seattle, Washington, July, 1972.
12. Preston, A., D. F. Jefferies, J. W. R. Dutton, B. R.
    Harvey, and A. K. Steele. *Environ. Poll., 3,* 69 (1972).
13. Noakes, J. E. and D. W. Hood. *Deep-Sea Res., 8,* 121
    (1961).
14. Armstrong, F. A. J., P. M. Williams, and J. D. H.
    Strickland. *Nature, 211,* 481 (1966).
15. Burrell, D. C. Proc. 3rd Intern. Atom. Spectrosc.
    Cong., Paris, August, 1972, pp. 409-428.
16. Slowey, J. F. and D. W. Hood. *Geochim. Cosmochim. Acta,
    35,* 121 (1971).
17. Brooks, R. R., B. J. Presley, and I. R. Kaplan. *Talanta,
    14,* 809 (1967).
18. Burrell, D. C., G. G. Wood, and P. J. Kinney. *Trans. Am.
    Geophys. Union, 49,* 759 (1968).
19. Davis, D. R. Ph.D. Thesis, Graduate College, Texas A & M
    University, College Station, Texas, 1968.

20. Whitnack, G. C. and R. Sasselli. *Anal. Chim. Acta, 47,* 267 (1969).

21. Spencer, D. W. and P. G. Brewer. *Geochim. Cosmochim. Acta, 33,* 325 (1969).

22. Zirino, A. and M. L. Healy. *Environ. Sci. Technol., 6,* 243 (1972).

23. Brewer, P. G. and D. W. Spencer. Geological Society of America Annual Meeting, Milwaukee, Wisconsin, November, 1970.

24. Burrell, D. C., V. M. Williamson and M.-L. Lee. 4th Intern. Conf. Atom. Spectrosc., Toronto, November, 1973.

25. Christian, G. D. 4th Intern. Conf. Atom. Spectrosc., Toronto, November, 1973.

26. Ihnat, M. 4th Intern. Conf. Atom. Spectrosc., Toronto, November, 1973.

27. Morgan, J. Institute of Marine Science Symposium on Organic Matter in Natural Waters, University of Alaska, College, Alaska, September, 1968.

28. Barsdate. in *Proc. of Symposium on Organic Matter in Natural Waters* (D. W. Hood, ed.), Institute of Marine Science Occasional Publ. No. 1, University of Alaska, 1970, pp. 485-493.

29. Gjessing, E. and G. F. Lee. *Environ. Sci. Technol., 1,* 631 (1967).

30. Slowey, J. F. Ph.D. Thesis, Graduate College, Texas A & M University, College Station, Texas, 1966.

31. Slowey, J. F., L. M. Jeffry, and D. W. Hood. *Nature, 214,* 377 (1967).

32. Longerich, L. L. and D. W. Hood. in *Clay-Inorganic and Organic-Inorganic Associations in Aquatic Environments* (D. C. Burrell and D. W. Hood, eds.), Institute of Marine Science Report R69-10(I), University of Alaska, 1968, pp. 1-11.

33. Mislar, J. P. and S. Elchuk. *Atomic Absorption Newsletter, 7,* 71 (1968).

34. Morrison, G. H. and H. Freiser. *Solvent Extraction in Analytical Chemistry,* John Wiley and Sons, New York, 1957.

35. Diamond, R. M. and D. F. Tuck. *Progr. Inorg. Chem., 2,* 109 (1960).

36. Irving, H. and R. J. P. Williams. in *Treatise on Analytical Chemistry* (I. M. Kolthoff and P. J. Elving, eds.), Vol. 3, John Wiley and Sons, New York, 1961, p. 1309.

37. Stary, J. *The Solvent Extraction of Metal Chelates,* MacMillan, New York, 1964.

38. Schweitzer, G. K. and W. van Willis. *Advan. Anal. Chem. Instr., 5,* 169 (1966).

39. De, A. K., S. M. Khopkar, and R. A. Chalmers. *Solvent Extraction of Metals,* Van Nostrand Reinhold, London, 1970.

40. Lee, M.-L. and D. C. Burrell. *Anal. Chim. Acta, 62,* 153 (1972).

41. Goldberg, M. C., L. deLong and L. Kahn. *Environ. Sci. Technol.*, *5*, 161 (1971).

42. Brooks, R. R. *Talanta*, *12*, 505 (1965).

43. Brooks, R. R. *Talanta*, *12*, 511 (1965).

44. Burrell, D. C. *Atomic Absorption Newsletter*, *7*, 65 (1968).

45. Pinta, M. *Detection and Determination of Trace Elements* (English Transl. from French by M. Bivas), Ann Arbor Science Publishers, Ann Arbor, Michigan, 1971.

46. Sandell, E. B. *Colorimetric Determination of Traces of Metals*, 3rd ed., John Wiley and Sons, New York, 1958.

47. Delaughter, B. *Atomic Absorption Newsletter*, *4*, 273 (1965).

48. Yanagisawa, M., M. Suzuki, and T. Takeuchi. *Anal. Chim. Acta*, *43*, 500 (1968).

49. Bode, H. and F. Neumann. *Z. Anal. Chem.*, *30*, 404 (1958).

50. Joyner, T., M. L. Healy, D. Chakravarti, and T. Koyanagi. *Environ. Sci. Technol.*, *1*, 417 (1967).

51. Ramakrishna, R. S. and H. Irving. *Anal. Chim. Acta*, *48*, 251 (1969).

52. Frei, R. W., G. H. Jamro, and O. Navratil. *Anal. Chim. Acta*, *55*, 125 (1971).

53. Dahl, I. *Anal. Chim. Acta*, *41*, 9 (1968).

54. Fritz, J. S., R. K. Gillette, and H. E. Mishmash. *Anal. Chem.*, *38*, 1869 (1966).

55. Fishman, M. J. and M. R. Midgett. in *Trace Inorganics in Water* (R. A. Baker, ed.), Advan. Chem. Ser., No. 73, American Chemical Society, Washington, D.C., 1968.

56. Burrell, D. C. *Anal. Chim. Acta*, *38*, 447 (1967).

57. Mansell, R. E. and H. W. Emmel. *Atomic Absorption Newsletter*, *4*, 365 (1965).

58. Jones, J. L. and R. D. Eddy. *Anal. Chim. Acta*, *43*, 165 (1968).

59. Takeuchi, T., M. Suzuki, and M. Yanagisawa. *Anal. Chim. Acta*, *36*, 258 (1966).

60. Sachdev, S. L. and P. W. West. *Anal. Chim. Acta*, *44*, 301 (1969).

61. Sachdev, S. L. and P. W. West. *Environ. Sci. Technol.*, *4*, 749 (1970).

62. Wakahayashi, T., S. Oki, T. Omori, and N. Suzuki. *J. Inorg. Nuc. Chem.*, *26*, 2255 (1964).

63. Alimarin, I. P. and Y. A. Zolotov. *Talanta*, *9*, 891 (1962).

64. Law, S. L.. *Atomic Absorption Newsletter*, *10*, 75 (1971).

65. Robinson, J. W., P. F. Lott, and A. J. Barnard. in *Chelates in Analytical Chemistry*, Vol. 4 (H. A. Flaschka and A. J. Barnard, eds.), Marcel Dekker, New York, in press.

66. Samuelson, O. *Ion Exchange Separations in Analytical Chemistry*. John Wiley and Sons, New York, 1963.

67. Helfferich, F. *Ion Exchange*. McGraw-Hill, New York, 1962.
68. Schmuckler, G. *Talanta, 12,* 281 (1965).
69. Biechler, D. G. *Anal. Chem., 37,* 1054 (1965).
70. Riley, J. P. and D. Taylor. *Anal. Chim. Acta, 42,* 548 (1968).
71. Ali, S. A. and D. C. Burrell. *Pakistan J. Sci. Ind. Res., 12,* 506 (1970).
72. Carritt, D. E. *Anal. Chem., 25,* 1927 (1953).
73. Chao, T. T., M. J. Fishman, and J. W. Ball. *Anal. Chim. Acta, 47,* 189 (1969).
74. Muzzarelli, R. A. A. *Talanta, 18,* 853 (1971).
75. Dingman, J., S. Siggia, C. Barton, and K. B. Hiscock. *Anal. Chem., 44,* 1351 (1972).
76. Baumann, A. J., H. H. Weetail, and N. Weliky. *Anal. Chem., 39,* 932 (1967).
77. Tera, F., R. R. Ruch, and G. H. Morrison. *Anal. Chem., 37,* 358 (1965).
78. Ruch, R. R., F. Tera, and G. H. Morrison. *Anal. Chem., 37,* 1565 (1965).
79. Minczewski, J. in *U.S. National Bureau of Standards Monograph 100,* (W. W. Meinke and B. F. Scribner, eds.), U.S. Gov't. Printing Office, Washington, D.C., 1967, pp. 385-416.
80. Simon, J., W. Schulze, and R. Reinke. *Z. Anal. Chem., 264,* 4 (1973).
81. Kim, Y. S. and H. Zeitlin. *Anal. Chim. Acta, 46,* 1 (1969).
82. Burke, K. E. *Anal. Chem., 42,* 1536 (1970).
83. Forster, W. and H. Zeitlin. *Anal. Chem., 38,* 649 (1966).
84. Biskupsky, V. S. *Anal. Chim. Acta, 46,* 149 (1969).
85. Heggen, G. E. and L. W. Strock. *Anal. Chem., 25,* 859 (1953).
86. Silvey, W. D. and R. Brennan. *Anal. Chem., 34,* 784 (1962).
87. Lai, M. G. and H. V. Weiss. *Anal. Chem., 34,* 1012 (1962).
88. Mitchell, R. L. and R. O. Scott. *Spectrochim. Acta, 3,* 367 (1948).
89. Mallory, E. C. in *Trace Inorganics in Water* (R. A. Baker, ed.), Advan. Chem. Ser., No. 73, American Chemical Society, Washington, D.C., 1968.
90. Riley, J. P. and G. Topping. *Anal. Chim. Acta, 44,* 234 (1969).
91. Tappmeyer, W. P. and E. E. Pickett. *Anal. Chem., 34,* 1709 (1962).
92. Firsching, F. H. in *Chelates in Analytical Chemistry,* Vol. 2 (H. A. Flaschka and A. J. Barnard, eds.), Marcel Dekker, New York, 1971.
93. Weiss, H. V. and J. A. Reed. *J. Marine Res., 18,* 185 (1960).
94. Weiss, H. V. and M. G. Lai. *Anal. Chim. Acta, 25,* 550 (1961).

CHAPTER 6

RADIATION SOURCES AND DETECTION

"The function of the optical train differs somewhat for the different models: flame emission and
atomic absorption.   However, the general arrangement
of an atomic absorption spectrometer is no different
from a flame emission spectrometer except for the
addition of a light source..."

J. A. Dean[1]

SOURCES

Spectral sources are required for both AA and AF
spectrometry. The design and correct choice of these
devices is exceedingly important for trace AS
analysis--as is optimization of all the instrumental
components--but this topic will not be pursued here
in very much detail since the analyst is basically
tied to available commercial equipment, for which a
voluminous body of literature exists. Most sources
depend for their operation on vaporization of the
free metal atoms followed by a dc or r.f. discharge.
It was shown in Chapter 3 that the source requirements for AA and AF differ in certain respects.
For the most sensitive absorption measurements a
source line considerably narrower than the absorption line is required (Figure 3.2); these widths are
determined by various line broadening processes,
among which Doppler and pressure broadening predominate. Source discharges are designed to occur
at very low pressures so that Lorentz and resonance
effects are negligible, and the width ratio is
effectively controlled by the relative temperatures.
This is tacitly assumed in Chapter 3 where the
theoretical derivatives are based on a Doppler

absorption half-width. One other line broadening
problem--that of self-absorption--is a potential
problem with all sources, but is most apparent with
vapor discharge designs as noted below.

The source requirements for AF are basically
similar to those for AA except that *intensity* is the
major consideration over line width. It has been
shown (Equation 3.14) that the fluorescence signal
is directly dependent upon the source intensity, so
that considerable effort has been directed toward
the development of high-intensity sources for this
analysis mode.

Continua

The use of continua as light sources has been
considered primarily on theoretical grounds in
Chapter 3. Such sources offer the possibility of
increased operating convenience and reduced costs
(compared with the maintenance of individual elemental sources), but for natural water analysis these
are not felt to be important considerations. Trace
analysis by AS necessitates operation of equipment
at the limits of detection capability, and continua
do not at present yield sufficiently intense sources,
regardless of selectivity requirements. However,
it is important to be familiar with the characteristics of these devices for one important reason,
namely, that they presently offer the potential for
automated AF analysis and, one day, this may be
extended also to AA systems.

The detector in AF analysis does not "see" the
source. Continuum sources have long formed the basis
of viable techniques,[2] although the detection limit
listing of Figure 3.6 clearly demonstrates an un-
favorable comparison with the line-source data.
More recently, there has been an unexpected resurgent
interest in the possible use of continuum sources
for AA. This is a quite different concept from the
AF application since simple calculation might suggest
that monochromators of exceedingly small band-pass
would be a requirement in the former case. However,
one must be on guard for the consequences of the
limiting assumptions inherent in the relevant basic
theory. De Galan *et al.*[3] have derived the following
expression for the ratio of the absorbances from line
and continuum sources:

$$\frac{A_{line}}{A_{cont.}} = \text{Const.} \ \bar{K} \cdot \frac{s}{\Delta\lambda_D} \tag{6.1}$$

In this equation the spectral band width-absorption line width term is all important. The absorption line width at conventional temperatures should be somewhere in the range 0.01-0.03 Å, and high resolution monochromators might be usable in certain cases. In fact, in recent years, several studies[4] have demonstrated surprisingly high sensitivities for continuum source AA, and de Galen and Samaey[5] have noted that, with a spectral band width of 0.18 Å, continuum source sensitivities are less than an order of magnitude worse than the equivalent line-source values. It was pointed out earlier that such results might not be unexpected for elements exhibiting complex spectra and hyperfine structure.

Xenon arcs have been the preferred sources for this type of work,[6] although certain instability problems have been recorded.[7] For the AA detection of metals in complex matrices, interference problems,[8] as expected, will undoubtedly complicate the use of continuum sources, and it is unlikely that AA automated analysis will rival the AF systems currently being developed.

Hollow Cathode Lamps

Hollow cathode (HC) lamps have been by far the most extensively utilized sources for AA analysis, and were initially advocated by Walsh[9] on the basis of their relatively intense but essentially narrow spectral-line characteristics. The latter feature has been the subject of some controversy, but measurements by Bruce and Hannaford[10] have shown that, for example, the calcium half-width ranges from 0.0092 to 0.0156 Å for lamp currents of 5-15 mA. This is in close agreement with the expected Doppler widths over the relevant temperature range of 347-429° K.

The currently available designs have been vastly improved over those available to the pioneers in the AA field, and intense competition among a number of manufacturers should ensure that this trend continues. Figure 6.1(a) shows the electrode arrangement for the basic HC lamp. The various designs, refinements and modes of operation of this source have been described in considerable detail in the basic texts.[7,11] The analyst has very little control over the design (multielement lamps are not recommended for trace analysis) or component arrangement of conventional lamps, but it may be found that use of a demountable HC lamp offers useful flexibility in certain cases.

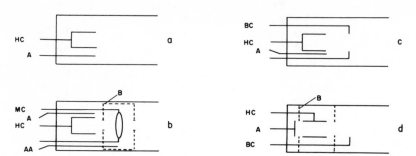

Figure 6.1.   Electrode arrangements of various HC lamp designs:
A - anode; HC - cathode; AA - auxiliary anode;
MC - modulator cathode; BC - booster cathode;
B - baffles
a Basic lamp design.
b HC lamp incorporating selective modulation.
(after Reference 12).
c High-intensity lamp.
d Improved high-intensity lamp design after
Reference 13.

For example, the filler gas employed may be chosen
to complement use of a particular resonance line.
Neon is the filler gas generally preferred for most
operations since it has a higher ionization potential and hence yields more intense lines than, for
example, argon.   Figure 6.2 illustrates one commercially available demountable design which is
approximately equivalent to a conventional high-
intensity HC lamp in output intensity.   Part of the
increased intensity has been attributed to a
suppression of self-reversal effects, but, as noted
elsewhere, such effects are not prominent for any
HC design.   These devices have also been successfully
used as AF spectral sources.[14]
High-intensity HC lamps were introduced some
years ago[15] for several elements (notably nickel).
These sources [Figure 6.1(c)] utilize a supplementary
and independent discharge operated at a sufficiently
low voltage to excite the primary atom cloud without
inducing ionization.   In the absence of readily
available electrodeless discharge tubes these lamps
are currently the most commonly chosen source for
AF analysis, but their utility in routine AA work
has, to a large extent, been supplanted by recent
improvements in basic HC design.   High-intensity

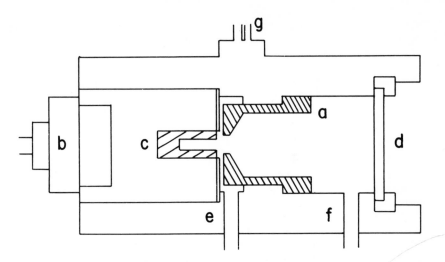

Figure 6.2. *Schematic of Glomax demountable HC lamp (by courtesy of Barnes Engineering Company)*

a*Anode*
b*Heat exchanger*
c*Hollow cathode*
d*Window*
e*Vacuum port*
f*Filler gas port*
g*Electrical connections*

lamps are considerably more expensive than the latter and necessitate a more complex power supply. Lowe[13] has recently introduced an improved high-intensity design [Figure 6.1(d)] which has shown marked improvement for AF work.

One final variation of the HC design worthy of note incorporates selective modulation for one particular metal[12] so that the lamp is capable of yielding the ac resonance emission usually accomplished by external modulation. The electrode arrangement for this lamp is shown as Figure 6.1(b). Source modulation is considered again below.

Recent improvements in conventional HC lamps have been concerned most usually with electrode design-- the chemical form of the cathodic element and electronic shielding, for example. The characteristics of the common filler gases--neon and argon--have been exhaustively discussed elsewhere.[7,16] The life expectancy of these lamps should be of great practical concern to the analyst since frequently decay may go

unnoticed on a day-to-day basis. It is good practice
to retain an initial scan record of each lamp to
facilitate later comparisons. Failure may be related
to loss of cathodic material, to incomplete expulsion
of air or subsequent leakage, and to adsorption of
the filler gas onto various internal surfaces.

Electrodeless Discharge Tubes

    If a thin glass or quartz tube containing inert
gas at low pressure and a small quantity of a metal
compound is subjected to an intense radio or microwave field, line emission occurs due to bombardment
of the metal within the alternating electromagnetic
field. These electrodeless discharge tubes (EDT)
give very intense and narrow lines, and are hence
eminently suitable for use as AF sources.[16-18] The
principal practical problem encountered is that of
coupling the electrical discharge to the sealed
vapor. The efficiency of this type of lamp is
largely dependent upon the choice of discharge
cavity employed; the sealed tube itself is relatively
simple to manufacture. The cavity is designed to
resonate and reflect maximum energy into the EDT,
and this is accomplished by tuning the frequency of
the cavity to the source so that impedance is matched
and a minimum of power is back-reflected. One such
cavity is illustrated in Figure 6.3. Medical dia-
thermy units have been shown to be useful microwave
excitation sources and commercial packages are now
available from one or two manufacturers. The best
of these incorporate reflected power meters to protect the electromagnetic source unit (and to aid in
maximizing emission). Tuning of the resonance
cavity for maximum spectral output and reproducibility
requires a fair degree of skill and practice, however,
and each individual tube may require quite different
operating conditions[19] for optimum results. It
appears to be preferable to operate the tubes at
moderately elevated temperatures; but it is more
important to maintain this temperature within narrow
limits since minor fluctuations drastically affect
the output intensity. Thermostating can be accomplished quite simply by, for example, blowing hot
air over the equipment. Cooke *et al.*[20] have given
a comprehensive critique of the EDT operating
parameters which require special attention. In the
past, most water quality laboratories have tended
to shy away from these sources (and hence from AF
analysis) for routine use, but this situation is

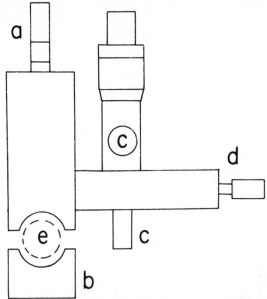

Figure 6.3. *Diagrammatic representation of a cavity (No. 214L, Electro-Medical Supplies Ltd.) for use with electrodeless discharge tubes. (After Reference 17 by courtesy of the Optical Society of America.)*
[a]*Tuning control*
[b]*Removable cap*
[c]*Cooling air inlets*
[d]*Coupling to microwave generator*
[e]*Position of discharge tube.*

changing and one manufacturer now offers a relatively simple system for arsenic and selenium.

Other Spectral Sources

Vapor discharge lamps (*e.g.*, Osram, Philips) also have small quantities of a particular metal within a sealed tube containing inert gas. The metal is vaporized and excited by a dc discharge between electrodes as shown in Figure 6.4. These lamps have long been used as AA and AF sources for the alkali elements, but less so now with the introduction of advanced HC lamp designs. The light from these lamps is very intense but--as a generalization--self-absorption effects are very pronounced and this lessens their attraction as AA sources.

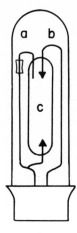

Figure 6.4.   Vapor discharge lamp.
a Starting electrode
b Main electrode
c Specific metal vapor

Self-absorption is, simply, AA occurring within the
lamp (Figure 3.2); in the extreme case self-reversal
of the line may occur.

Several other sources for AA and AF analysis have
been advanced, including direct metallic sputtering[21]
and sparks.[22]   Until recently, the advocated advantages of these devices have been insufficiently well
established to compensate for obvious disadvantages
such as cost and limited applicability in the aqueous
trace metal laboratory.   The current availability of
tunable lasers,[23,24] however, now offers for the
first time the tantalizing prospect of a practical
AF analysis method for a wide range of metals present
in solution in trace and ultratrace amounts.[25]
Winefordner[26] has recently presented AF relative
aqueous detection limits in the ng-µg/ℓ range for a
suite of metals (including chromium, molybdenum,
titanium and vanadium) *via* a continuous sampling
mode using the pulsed, tunable dye laser.   As noted
in Chapter 3, such ultra-intense devices are
saturation-radiation sources.[27]   In practical terms
this means that, since the AF signal is no longer a
function of the source intensity, minor fluctuations
of the latter are unimportant and, additionally, the
specific flame chemistry used for atomization is of
little consequence.

OPTICAL SYSTEM

Choice of Spectral Line

For most routine AA and AF work, the analyst will not generally experience too much difficulty selecting a suitable source wavelength, since copious "standard operating condition" data are freely available in atlas form,[28] within the standard texts, and from instrument and lamp manufacturers. This volume, however, is primarily addressed to the analyst concerned with high precision trace analysis. Some understanding of the principles and problems involved should allow needed flexibility and should generally enhance the power and utility of these AS methods.

Absorption Analysis Lines

It has been shown (Equation 3.9) that, at a given frequency, the absorption coefficient is a function of N_O and f, the oscillator strength of the transition (as a simplification, the lower energy state has been taken to be the ground state; this is reasonable for the temperature ranges usually encountered). Listed f values are therefore a valuable guide to the value of specific absorbing lines, where other factors are not limiting. Table 6.1 lists the major resonance lines used for AA analysis together with a selection of less sensitive lines. The corresponding f value data[29] should be noted. Margoshes[30] has considered in some detail all the relevant AA line selection criteria. The sensitivity data included in Table 6.1 are based upon the common ppm/1% absorption definition discussed in Chapter 9. Absorbing lines of varying sensitivity may also be selected from a comparison of standard and source spectral scans. A few of the otherwise more suitable resonance lines lie in the far UV region and are either inaccessible to conventional equipment or are difficult to use, principally because of flame opaqueness. This has led to the development of the special flames (*e.g.*, for Se and As) considered in Chapter 7.

Fluorescence Analysis Lines

The most favored lines for AF analysis are commonly the principal absorption lines given above,

Table 6.1

Atomic Absorption Analysis Lines

Element	Major Lines[a] (Å)	Minor Lines[b] (Å)	f Values[c]	Sensitivity[d] Reference 29	Reference 31	Flame
Ag	3281		0.51	0.05	0.05	1
		3383	0.25	0.15		1
As	1937		0.10	5.00	1.50	1,3
		1972	0.07			1
Au	2428		0.30	0.30	0.30	1
		2676	0.19	1.30		1
Bi	2231		0.01	0.70	0.70	1
		2229		1.50		1
Cd	2288		1.20	0.03	0.04	1
		3261		20.00		1
Co	2407		0.22	0.16	0.20	1
		2425	0.19	0.40		1
Cr	3579		0.34	0.15	0.10	1
		4254	0.10	0.50		1
Cu	3248		0.74	0.10	0.50	1
		3274	0.38	0.20		1
Fe	2483[e]		0.34	0.14	0.10	1
		2523	0.30	0.20		1
		3720	0.04	1.00		1
Ga	2874		0.32	3.00	2.00	1
		2944	0.29	4.00		1
Ge	2651		0.84	2.50	2.50	2
Hf	3073		0.09	14.00		2
Hg	2537		0.03	10.00		1
In	3039		0.36	1.00		1
		3256	0.37	1.00	1.00	1
Ir	2640[f]		0.06	44.00	20.00	1

Table 6.1, continued

Element	Major Lines[a] (Å)	Minor Lines[b] (Å)	f Values[c]	Sensitivity[d] Reference 29	Reference 31	Flame
Mn	2795[g]		0.58	0.05	0.05	1
		2801	0.29	0.15		1
		4031	0.06	1.00		1
Mo	3133		0.20	1.70		1
		3170	0.12	40.00		2
Nb	3349		0.09	40.00		2
		4059	0.19	35.00		2
Ni	2320[h]		0.09	0.2		1
		2310	0.07	0.4	0.1	1
		3415	0.30	3.0		1
Os	2909			3.0	2.0	2
		3059		5.0		2
Pb	2170		0.40	0.5	0.5	1
		2833	0.21	1.0		1
Pd	2476		0.10	0.8	0.5	1
		2448	0.07	1.0		1
Pt	2659		0.12	2.0	3.0	1
Re	3460		0.20	20.0		2
Rh	3435		0.07	0.3	1.0	1
	3692		0.06	0.6		1
Ru	3499		0.10	2.0	1.5	1
Sb	2176		0.05	0.5	0.7	1
		2068	0.10	1.0		1
		2311	0.03	1.5		1
Se	1960		0.12	3.0	0.4	1,3
		2040	0.26	5.0		1
Sn	2246		0.41	0.7	0.4	3
		2863	0.23	1.3		3
Ta	2715		0.06	30.0		2

Table 6.1, continued

Element	Major Lines[a] (Å)	Minor Lines[b] (Å)	ƒ Values[c]	Sensitivity[d] Reference 29	Reference 31	Flame
Te	2143		0.08	1.5		1
Ti	3643		0.25	1.4	3.0	2
		3654	0.22	1.6		2
Tl	2768		0.27	0.8		1
		2380	0.07	2.0		1
			0.66			
V	3184[g]			1.3	2.0	2
		3066		0.5		2
		3704		0.5		2
W	4009			35.0		2
Zn	2139		1.2	0.02	0.02	1
Zr	3601		0.22	20.0		2

Flames: 1 - Air-acetylene
         2 - Nitrous oxide-acetylene
         3 - Hydrogen-air

[a] Most sensitive or otherwise recommended analytical line.

[b] Other absorbing lines.

[c] Oscillator strength data from Reference 29.

[d] Relative sensitivity data from references cited in ppm/1% absorption units (see text).

[e] Doublet with less absorbing line at 2488 Å.

[f] Unresolved doublet.

[g] Triplet.

[h] To be distinguished from ion line at 2316 and less absorbing line at 2311 Å.

since resonance fluorescence--which results from
excitation and deactivation from and to the ground
state (Figure 3.3)--generally yields optimum sensitivity. Other deactivation modes may offer analytical
advantages in certain cases however, so that a
multiple choice between a number of emission lines
exists for each incident line. Bailey[32] has outlined
a theoretical approach to this problem which may aid
in selecting a suitable line for a given set of
experimental conditions. The major AF analytical
lines used with both line and continuum sources are
included in Table 6.2.

Table 6.2

Atomic Fluorescence Analysis Lines

Element	Line Source	Continuum	Element	Line Source	Continuum
Ag		3281	Os		-
As	1937		Pb		4058
Au		2676	Pd		3405
Bi	2231	3068	Pt		-
Cd		2288	Re		-
Co		2407	Rh		3692
Cr	3572	3579	Ru		-
Cu		3248	Sb	2311	
Fe		2483	Se	1960	
Ga		4172	Sn	3034	
Ge	2652		Ta		-
Hf	2866		Te	2143	
Hg	2537		Ti	3200	
In	4511	4105	Tl		3776
Ir		2544	V	3184	
Mn		2795	W		-
Mo	3133		Zn		2139
Nb		-	Zr		-
Ni		2320			

Emission Analysis Lines

Selection of a suitable line for emission analysis is a considerably more skillful operation than for AA or AF since the choice must take account of potential spectral interference (see Table 8.1) and a reasonable familiarity with the sample matrix is required. At the same time the choice of lines for each element is enhanced since this analysis mode is not limited to ground-state atomic transitions, and various bands and ion emissions may also be utilized. Table 6.3 lists suitable analysis lines for flame emission analysis. These wavelengths should be considered in conjunction with the information given in Table 8.1 (and the more comprehensive listings included in the standard texts, *e.g.*, Reference 33) concerning potentially interfering concomitants.

Table 6.3

Flame Emission Analysis Lines

Element	Major Lines[a] (Å)	Minor Lines[b] (Å)	Radiation Species	Excitation Energy (ev)	Sensitivity[c] Reference 33	Flame
Ag	3281		Ag	3.78	1.0	2
		3383	Ag	3.66	0.6	2
As	–	2350	As	6.60		
Au	2676		Au	4.63		
Bi	–	2231	Bi	5.5		
		3068	Bi	4.04		
Cd	3261		Cd	3.80	0.5	2
		2288	Cd	5.41	10.0	3
Co	3454		Co	4.02	3.4	1
		2425	Co	5.10		
		3527	Co	3.51	4.0	1
Cr	4254		Cr	2.91		
		3605	Cr	3.43		
Cu	3248		Cu	3.82		
		3274	Cu	3.78		

Table 6.3, continued

Element	Major Lines[a] (Å)	Minor Lines[b] (Å)	Radiation Species	Excitation Energy (ev)	Sensitivity[c] Reference 33	Flame
Fe	3720		Fe	3.32	2.5	1
		2483	Fe	4.99		
		3860	Fe	3.21	2.7	1
Ga	4172		Ga	3.07	0.5	1
		4033	Ga	3.07	1.0	1
Ge	2652		Ge			
Hg	2537		Hg	4.88	10.0	3
In	4511		In	3.02	0.1	2
		4102	In	3.02	0.1	2
Mn	4031		Mn	3.08	0.1	1
		559(0)	MnO		2.7	1
Mo	3798			3.26		
		3133				
		3194				
Nb	4059		Nb			
Ni	3415		Ni	3.65	3.5	1
		3525	Ni	3.54	1.6	1
Pb	4058		Pb	4.38	14.0	1
		3684	Pb	4.34	21.0	1
Pd	3635		Pd	4.23	0.1	2
		3405	Pd	4.46	0.2	2
Pt	-	2659	Pt	4.66		
Re	3461		Re	3.58		
Rh	3692		Rh	3.35	0.7	2
		3435	Rh	3.60	2.0	2
Ru	3728		Ru	3.32	0.3	1
Sb	2598		Sb			
		2312	Sb	-		
		2528	Sb	6.12		

Table 6.3, continued

Element	Major Lines[a] (Å)	Minor Lines[b] (Å)	Radiation Species	Excitation Energy (ev)	Sensitivity[c] Reference 33	Flame
Se	2040		Se			
Sn	2840		Sn			
		2355	Sn			
		2430	Sn	5.1		
Te	2143		Te			
		2386	Te	5.80		
Ti	3999		Ti	3.15		
		3656	Ti			
		517 (O)	TiO		10.0	2
Tl	3776		Tl		0.6	2
		3531	Tl		1.2	2
V	3184		V			
		4112	V			
		4379	V	3.13		
		523 (O)	VO		12.0	2
W	4009		W	3.09		
Zn	2139		Zn	5.8	500.0	1
	3076		Zn	4.03		
Zr	3520		Zr			
		3601	Zr			

Flames: 1 - Oxyacetylene
          2 - Oxyhydrogen
          3 - Air-hydrogen

[a] Most sensitive or otherwise recommended analysis line.

[b] Other lines which have been utilized for flame emission analysis.

[c] Compilation of relative sensitivity data from Reference 33 using the combustion mixtures cited; ppm/1% transmission units.

Beam Optics

A diagrammatic representation of the optical
train of an idealized AS system is included in
Figure 3.6. Every available commercial spectrometer
package is, of course, unique in some respect in the
precise arrangement of these individual components.
A thoroughly detailed background to this subject has
been given by Dean[1] and only a few areas of particular concern to high precision trace analysis will be
emphasized here.

The characteristics of the absorption cell--
defined by the geometry of the source beam through
the flame--are of paramount importance to AA and AF
analysis. As a generalization, the maximum length
of flame must be traversed by the incident light for
optimum absorption. The most efficient lens arrangement is that which passes an essentially parallel
beam through the burner; most systems converge the
beam at the center of the cell. It is most important
to adjust the beam position so as to intercept the
maximum volume of the requisite flame zone as discussed in the following chapter. Since no absorption
can occur outside the cell, it is a useful exercise
to chart the source beam through the flame region
using, where necessary, the visible filler-gas
emission. In this fashion, unnecessary dilution
of the sample due to injection into noncell regions
may be avoided. Some AS instruments incorporate
an optical multipass system through the flame to
effectively increase the cell length, but in this
case the analyst must be especially aware of potential flame effects--spurious emissions and irregular
fluctuations of the cell geometry. Flame flicker
is, in fact, a major factor controlling the precision
limits of any flame analytical procedure. Currently
available commercial double-beam equipment obviously
cannot compensate for such flame effects since only
the analysis beam traverses the flame. Stephens,[34]
however, has noted that Zeeman splitting of the
source line can be used to produce a true, compensating "double-beam" system. By this device, the
perturbed and non-perturbed Zeeman components act
as the reference and sample beams respectively, and
both are equally subject to undesirable modifications
within the flame.

One other optical phenomenon worth some consideration concerns the optical distortion of the source
beam by the flame. Boling[35] has suggested that the
superior performance of the 3-slot ("Boling") burner

may be attributed to a reduction of these effects. Certainly some such explanation is indicated since the design theoretically should suffer from enhanced sample dilution. Unit (3) of Figure 3.6 refers to the continuum background correction source[36] which is recommended as an essential accessory for trace analysis, especially where complex and variable matrices are involved.

For AF analysis, the high intensity source must irradiate the required flame reaction zone as efficiently as possible. Thereafter the objective is as for flame emission, that is, to "gather" as large an emission area as possible at the monochromator slit.

The Monochromator

AA analysis lines need only to be separated from adjacent nonabsorbing lines of the test element (assuming the use of a line source). The complex spectra of the heavy metals will necessitate the use of something better than filters, of course, but for many applications the dispersion requirements are not stringent. This generalization applies equally or more so to AF analysis. In fact, if flame emission may be negated, it is possible--even desirable in some cases--to use nondispersive systems for AF (as considered briefly below). Conversely, flame AE analysis of the heavy metals may well be limited by the high dispersion required, particularly where coexisting matrix elements interfere. Similar considerations apply to continuum source AA and AF analysis. The use of extremely high resolution echelle gratings[37] has been noted previously.

ELECTRONIC SYSTEM

Modulation and Stability

The importance of a stabilized source for AA and AF has been referred to several times. Other factors being equal, source emission fluctuations will decrease the signal-to-noise ratio of the analysis. These effects may be alleviated by the use of double-beam instrumentation. However, modern lamp design and close attention to lamp current stabilization

design have largely dispelled the more serious
problems originally encountered with single-beam
operation.

All standard AS instruments incorporate the
facility for modulation of the source emission as
advocated in the pioneering Walsh AA treatise.[9]
In a modulated system, the source is a pulsed (ac)
beam, and the detection system is tuned to respond
only to the modulated frequency so that the continuous thermal emissions from the flame are rejected.
This type of correction is obviously denied to flame
emission analysis (although the latter emissions
must frequently be modulated also in order to be
compatible with the AA/AF detection system), so
that emission signals are superimposed upon the
background flame spectra. Modulation may be by two
methods--mechanical or electronic. Mechanical
modulation utilizes a rotating chopper placed in
front of the source beam; the chopper may also act
as a beam splitter in the case of double-beam instrumentation. The HC lamps of single-beam instruments are invariably electronically modulated and
frequency-tuned to the synchronized photometer
system. EDT sources may be modulated by means of
a mechanical chopper superimposed between source
and flame, but electronic modulation has also been
successfully applied[38] and the relative merits of
the two modes have been compared by Wildy and
Thompson.[39]

Walsh and co-workers[40,41] have advocated the use
of an ingenious selective modulation system whereby
the light from an unmodulated source is passed
through a pulsed atomic vapor of a metal generated
in a second lamp. Only the resonance lines of this
metal are thus modulated; all other lines are un-
affected and hence rejected by the tuned ac detection
system. Very large slit widths may be used and, in
certain favorable cases, the monochromator may be
entirely dispensed with. This basic system has been
subsequently refined[42] so that the pulsating atomic
vapor cloud is maintained within the primary source
lamp as illustrated in Figure 6.1(b).

Detection and Amplification

Photomultipliers are generally used as radiation
detectors for AS analysis[29,43] and Figure 6.5 shows
the wavelength response of several commonly employed
types. Instrumentation design for further signal

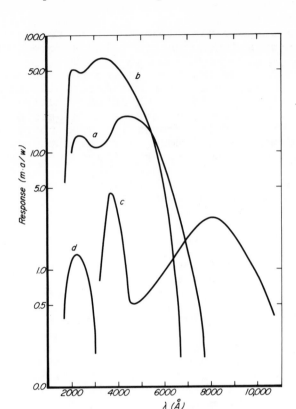

Figure 6.5.   Spectral sensitivity (response) curves for some
HTV (Hamamatsu TV Co. Ltd.) photomultiplier
detectors.

^{a}R213
^{b}R106
^{c}R196
^{d}R166 (a solar-blind type).

discrimination and amplification is well covered in
the standard texts (Reference 44, for example). It
has been noted above that frequently AA and AF signals need only be isolated in a fairly simple fashion
from the background flame emissions. Larkins *et al.*[45]
have thus suggested the general application of
solar-blind photomultipliers to these analysis modes.
Detectors of this type do not respond to radiation
signals at wavelengths in excess of around 3200 Å
(Figure 6.5), and most flame emission bands lie in
this forbidden region (Figure 6.6A). Solar-blind

detectors have been used[46],[47] in practical AF instrumentation arrangements in the absence of any
monochromator as illustrated in Figure 6.6B.

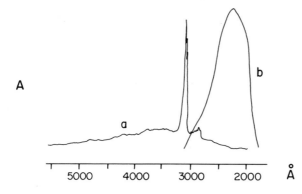

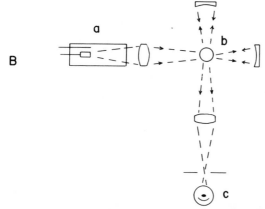

Figure 6.6. *Monochromatorless AF analysis by means of a*
solar-blind photomultiplier. *(After Reference*
12 by courtesy of the International Union of Pure
and Applied Chemistry.)
A. Spectral sensitivity curve of cesium-telluride
solar-blind photomultiplier R-166.

aAir-acetylene emission spectrum.
bR-166 sensitivity (response) curve.

B. Schematic of AF analysis equipment using a
solar-blind photomultiplier.

aHC lamp source.
bFlame absorption fluorescence cell.
cSolar-blind photomultiplier.

Resonance detectors[41],[48],[49] may also replace conventional optical monochromators. These devices (Figure 6.7) may be thought of as functioning as HC lamps in reverse. Resonance radiation impinges on the atomic vapor produced in the detector, is absorbed and re-emitted as a fluorescence signal which is measured in the usual fashion.

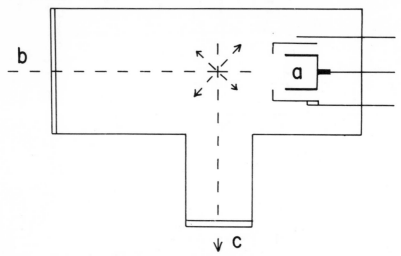

Figure 6.7. A resonance monochromator. (After Reference 29 by courtesy of Iliffe Books Ltd.)

a Hollow cathode
b Source radiation
c Fluorescence emission to detector.

It is stressed in Chapter 9 that the detection and read-out requirements for AS systems employing either continuous or discrete sample atomization are quite different. The instrumentation considered so far is especially applicable to the measurement of the continuous "equilibrium" signals, whereas discrete samples theoretically necessitate integration techniques. The possible application of direct photon counting to this latter type of AS analysis has recently been considered.[50-52] Photon counting is particularly suited to the detection of low-level radiation; that is, it is applicable to AF rather than to AA analysis since, in the latter case, relatively intense signals are being monitored. But the need for fast response-time detection and display systems for filament and furnace atomization will

remain for many a vexing problem so long as spectrometers specifically designed for this atomization
mode remain unavailable. A reasonably uncomplicated
dc system with oscilloscope display, for example,
has been suggested.[53]

Multi-element AS analysis has not been stressed
in these pages. The necessity of developing this
analytical capability for aqueous pollution monitoring
is well recognized, but much of the methodology and
instrumental components are logical extensions of
that required and described for single-element
analysis. Detection systems, however, present
innumerable problems, as might be expected, and one
ingenious solution[54] involves the application of
multi-channel vidicon (*i.e.*, TV camera tube) detectors. The necessary components must be custom
assembled at present but this should not present
insurmountable difficulties for those laboratories
handling large numbers of samples requiring the
routine determination of the same suite of elements.

Electronic Noise

The detection limit ranges of the AS methods
(see Chapters 3 and 7) are signal-to-noise ratios,[55]
and photomultiplier noise is an important factor
which could restrict the achievement of better detection capabilities. This is particularly so where
nonflame atomization cells are used. In such cases--
in which flame noise is no longer a major considera-
tion--it has been noted (*e.g.*, Reference 56) that AA
and AF are source-noise and phototube-noise limited
respectively. Photomultiplier shot noise is a
statistical uncertainty term due to the arrival of
photons as individual events. The rms shot-noise
current (I_s) is given by:

$$I_s = \text{Const} \cdot (I_{tot} \cdot \Delta F)^{1/2} \tag{6.2}$$

so that this noise effect is proportional to the
square of the primary current.

There is also a thermal noise to be considered
(see Reference 57) resulting from random motion of
the relevant changed particles. The thermal rms
noise (as a voltage E_t) is:

$$E_t = (4RkT\Delta F)^{1/2} \tag{6.3}$$

Thermal noise may thus be minimized by keeping both the resistance (R) and the temperature (T) of the photomultiplier circuits low. If these effects are reduced to a minimum, the electronic noise control on the detection limit will be due to the photomultiplier dark current. This is a background signal--in the absence of impinging radiation--resulting from operational current leakage.

REFERENCES

1. Dean, J. A. in *Flame Emission and Atomic Absorption Spectrometry*, Vol. 2 (J. A. Dean and T. C. Rains, eds.), Marcel Dekker, New York, 1971, pp. 111-147.
2. Veillon, C., J. M. Mansfield, M. L. Parsons, and J. D. Winefordner. *Anal. Chem., 38,* 204 (1966).
3. de Galan, L., W. W. McGee, and J. D. Winefordner. *Anal. Chim. Acta, 37,* 436 (1967).
4. McGee, W. W. and J. D. Winefordner. *Anal. Chim. Acta, 37,* 429 (1967).
5. de Galan, L. and G. F. Samaey. *Spectrochim. Acta, 25B,* 245 (1970).
6. Frank, C. W., W. G. Schrenk, and C. E. Meloan. *Anal. Chem., 39,* 534 (1967).
7. Butler, L. R. P. and J. A. Brink. in *Flame Emission and Atomic Absorption Spectrometry*, Vol. 2 (J. A. Dean and T. C. Rains, eds.), Marcel Dekker, New York, 1971, pp. 21-56.
8. West, T. S. in *Atomic Absorption Spectroscopy* (R. M. Dagnall and G. F. Kirkbright, eds.), Butterworths, London, 1970, pp. 99-126.
9. Walsh, A. *Spectrochim. Acta, 7,* 108 (1955).
10. Bruce, C. F. and P. Hannaford. *Spectrochim. Acta, 26B,* 207 (1971).
11. Burger, J. C., W. Gillies and G. K. Yamasaki. in *Analytical Flame Spectroscopy: Selected Topics* (R. Mavrodineanu, ed.), MacMillan, London, 1970, pp. 625-649.
12. Walsh, A. in *Atomic Absorption Spectroscopy* (R. M. Dagnall and G. F. Kirkbright, eds.), Butterworths, London, 1970, pp. 1-10.
13. Lowe, R. M. *Spectrochim. Acta, 26B,* 201 (1971).
14. Dinnin, J. I. and A. W. Helz. *Anal. Chem., 39,* 1489 (1967).
15. Sullivan, J. V. and A. Walsh. *Spectrochim. Acta, 21,* 721 (1965).
16. Mansfield, J. M., M. P. Bratzel, H. O. Norgordon, D. O. Knapp, K. E. Zacha, and J. D. Winefordner. *Spectrochim. Acta, 23B,* 389 (1968).

17. Dagnall, R. M. and T. S. West. *Appl. Opt.*, *7*, 1287 (1968).
18. Zacha, K. E., M. P. Bratzel, J. D. Winefordner, and J. M. Mansfield. *Analyt. Chem.*, *40*, 1733 (1968).
19. Silvester, M. D. and W. J. McCarthy. *Anal. Lett.*, *2*, 305 (1969).
20. Cooke, D. O., R. M. Dagnall, and T. S. West. *Anal. Chim. Acta*, *54*, 381 (1971).
21. Human, H. G. C. and L. R. P. Butler. *Spectrochim. Acta*, *35B*, 647 (1970).
22. Strasheim, A. and H. G. C. Human. *Spectrochim. Acta*, *23B*, 265 (1968).
23. Fraser, L. M. and J. D. Winefordner. *Anal. Chem.*, *43*, 1693 (1971).
24. Fraser, L. M. and J. D. Winefordner. *Anal. Chem.*, *44*, 1444 (1972).
25. Omenetto, N., N. N. Hatch, L. M. Fraser and J. D. Winefordner. *Spectrochim. Acta*, *28B*, 65 (1973).
26. Winefordner, J. D. 4th Intern. Conf. Atom. Spectros., Toronto, November 1973.
27. Piepmeier, E. H. *Spectrochim. Acta*, *27B*, 431 (1972).
28. Parsons, M. L. and P. M. McElfresh. *Flame Spectroscopy: Atlas of Spectral Lines*, Plenum, New York, 1971.
29. Rubeska, J. and B. Moldan. *Atomic Absorption Spectro-photometry* (English translation, P. T. Woods, ed.), Iliffe, London, 1969.
30. Margoshes, M. *Anal. Chem.*, *39*, 1093 (1967).
31. Reynolds, R. J. and K. Aldous. *Atomic Absorption Spectroscopy: A Practical Guide*, Griffin, London, 1970.
32. Bailey, B. W. *Spect. Lett.*, *2*, 81 (1969).
33. Dean, J. A. *Flame Photometry*, McGraw-Hill, New York, 1960.
34. Stephens, R. 4th Intern. Conf. Atom. Spectros., Toronto, November 1973.
35. Boling, E. A. *Spectrochim. Acta*, *23B*, 495 (1968).
36. Sandoz, D. P. and D. L. Murray. *Resonance Lines*, *2*, 7 (1970).
37. Cresser, M. S., P. N. Keliher, and C. Wohlers. *Spect. Lett.*, *3*, 179 (1970).
38. Browner, R. F., R. M. Dagnall, and T. S. West. *Anal. Chim. Acta*, *46*, 207 (1969).
39. Wildy, P. C. and K. C. Thompson. *Analyst*, *95*, 562 (1970).
40. Bowman, J. A., J. V. Sullivan, and A. Walsh. *Spectrochim. Acta*, *22*, 205 (1966).
41. Sullivan, J. V. and A. Walsh. *Appl. Opt.*, *7*, 1271 (1968).
42. Lowe, R. M. *Spectrochim. Acta*, *24B*, 191 (1969).
43. Dvorak, J., I, Rubeska, and Z. Rezac. *Flame Photometry: Laboratory Practice* (English translation, R. E. Hester, ed.), CRC Press, Cleveland, 1971.
44. Veillon, C. in *Flame Emission and Atomic Absorption Spectrometry*, Vol. 2 (J. A. Dean and T. C. Rains, eds.), Marcel Dekker, New York, 1971, pp. 149-176.

45. Larkins, P. L., R. M. Lowe, J. V. Sullivan, and A. Walsh. *Spectrochim. Acta, 24B,* 187 (1969).
46. Larkins, P. L. *Spectrochim. Acta, 26B,* 477 (1971).
47. Elser, R. C. and J. D. Winefordner. *Appl. Spectrosc.,* 25, 345 (1971).
48. Sullivan, J. V. and A. Walsh. *Spectrochim. Acta, 21,* 727 (1965).
49. Muller, R. M. *Anal. Chem., 39,* 99A (1967).
50. Cooke, D. O., R. M. Dagnall, B. L. Sharp, and T. S. West. *Spect. Lett., 4,* 91 (1971).
51. Alger, D., R. G. Anderson, I. S. Maines, and T. S. West. *Anal. Chim. Acta, 57,* 271 (1971).
52. Malmstadt, H. V., M. L. Franklin, and G. Horlick. *Anal. Chem., 44,* 63A, 1972.
53. Alger, D., R. G. Anderson, I. S. Maines, and T. S. West. *Anal. Chim. Acta, 57,* 271 (1971).
54. Busch, K. W. and G. H. Morrison. *Anal. Chem., 45,* 712A (1973).
55. Winefordner, J. D., W. J. McCarthy, and P. A. St. John. *J. Chem. Educ., 44,* 80 (1967).
56. Winefordner, J. D. and R. C. Elser. *Anal. Chem., 43,* 24A (1971).
57. Skogerboe, R. K. in *Flame Emission and Atomic Absorption Spectroscopy,* Vol. 1 (J. A. Dean and T. C. Rains, eds.), Marcel Dekker, New York, 1969, pp. 381-411.

CHAPTER 7

SAMPLE ATOMIZATION

"The most important component in flame spectrometry, whether emission or absorption, is the
nebulizer and burner system."
                              R. Hermann[1]

"Paradoxical though it may be ... the selection
of the best method of atomizing is a much more
complex problem in atomic absorption analysis than
in atomic emission analysis."
                              B. V. L'vov[2]

The theoretical basis of atomic emission, absorption and fluorescence has been covered in Chapter 3;
it is now necessary to consider the means available
for preparing atomic vapor in a practical fashion
so that the subsequent atomic transformations may
be monitored. This is a difficult but most important
operation in AS analysis, and particularly so for the
determination of metals present in limiting concen-
trations--presumably the prime concern of the reader.
Many texts are available covering these topics and
it would be redundant here to reiterate this material
in any detail. However, because optimization of the
atomization steps is vital to successful trace
analysis, and in view of the rapid strides being
made in this field, it seems necessary to review
these operations with emphasis on practical applications to natural water analysis.
    Throughout the following pages it must be remem-
bered that the requirements for AE and for AA/AF
analysis are quite different. For the latter, therma
energy is supplied only in order to provide a free
atomic environment for subsequent irradiation. But

159

emission analysis necessitates additional excitation
or ionization of the analyte atoms so that, by comparison, higher activation temperatures are needed.
It is for this reason that some consideration of
plasmas and other highly energetic sources has been
included, although, as will be seen, these *a priori*
energy requirement distinctions are frequently
blurred in practice. As a corollary to this, it
follows that undesirable temperature quenching effects are likely to accompany the addition of the
aqueous sample during AE analysis, whereas minor
temperature fluctuations during AA or AF analysis
are relatively unimportant. It should be noted also
that the flames and electrically heated devices
which are chiefly considered in this chapter are
only, for most people, the most readily available
and not necessarily the most efficient means of
atomizing the sample. The plasmas mentioned above
are briefly covered in a concluding section, but the
use of shock-wave excitation, for example, has also
been advocated for aqueous trace-metal analysis.[3]
Cathodic sputtering[4,5] is not, of course, directly
applicable to solution analysis, but atomization in
this fashion *via* direct ion impact avoids many of
the problems associated with thermal activation
(differential volatilization of sample species, for
example).

It is believed to be a useful concept to think of
absorption or emission of radiation as from a finite
cell, so that, for optimum sensitivity and detection,
it is necessary to arrange for the maximum concentration of atomic vapor within this cell. The boundaries
of such AS cells are defined--depending upon the mode
of analysis--by the source irradiation and light-
gathering properties of the spectrometer. Thus, for
AA analysis, only those atoms physically within the
cone or beam of the incident source radiation are of
analytical value and, in this case, the cone area
defines the cell. The same reasoning applies equally
to the absorption stage of AF, but since emissions
are omni-directional, the spatial configuration of
the AF and AE cell is of lesser importance.

Consideration of the enclosed nonflame atomization
units as atomic vapor cells is a relatively straightforward exercise, although the interior dimension of
particular furnace designs, for example, may not
exactly define the absorption volume since this
depends upon the geometry of the source beam. Flame
cells, on the other hand, are a considerably more
nebulous and complex concept (being anisotropic and
dynamic) and it is necessary to consider also the

transit rate of sample through specific flame reaction zones. Sample for furnace and filament atomizers, and also for total consumption burners (see below), is introduced--unmodified--directly into the cell, and the degree of atomization (β) is largely dependent upon dissociation and ionization reactions (Chapter 3); but also on a variety of condensed-phase transformations which, in the case of flames, must be completed very rapidly. By way of contrast, the most commonly utilized form of flame AA analysis makes use of a premix burner configuration in which the sample is initially taken up into a mixing chamber and subsequently undergoes a series of physical transformations *prior* to being sprayed into the flame in a form better suited to efficient free-atom formation within the cell. These basic AS cell patterns are considered further below.

Apart from one example of "impulse atomization" into a nitrous oxide-acetylene flame,[6] flame AS analysis requires a continuous flow of sample solution to the flame cell, and high precision analysis may necessitate the aspiration of a volume of sample in excess of 1 ml per element. Flameless atomization-- as presently utilized--consumes a single sample injection in the μℓ range, so that the final reliability of these data may be limited by errors inherent in reproducing these small volumes. Such "discrete" sampling analysis presents its own peculiar problems, and these also are covered in later sections of this chapter.

Both flame and nonflame AS methods are advantageously applicable to solution analysis since the analyte species are then in an homogeneous medium which may be reproducibly converted to the free atomic state within the cell. This book is, of course, primarily concerned with the analysis of liquid samples, but it should be appreciated that AS analysis is, under certain circumstances, applicable also to the direct determination of metals present in solid phases. For example, both Segar *et al.*[7] and Lord *et al.*[8] have analyzed biota tissue in furnace atomizers; dry aerosols are, as noted later, particularly advantageous for plasma excitation.[9] Various attempts[10] to directly analyze solid samples *via* flame atomization have, however, met with little success, although small quantities of particulate matter in aqueous solution may be accommodated.[11]

FLAMES AND BURNERS

Flames are the most common energy source and
until recently have been the most convenient medium
in which to measure the excitation/de-excitation AS
reactions considered in Chapter 3. The processes
which occur when the sample is fed to a flame cell
are, however, exceedingly complex and, since the
sample species pass through the critical flame zones
very rapidly, equilibrium conditions are exceptional
and it is not usually possible to characterize these
kinetically controlled reactions very closely.

The complete sequence of the transformation of
sample to free atomic vapor (nebulization, desolvation, vaporization/dissociation) has been illustrated
in a general fashion in Figure 3.2, and is discussed
in detail by, for example, Alkemade.[12] Nebulization
in premix chambers is considered below; sample
desolvation may subsequently be substantially completed in this type of burner *prior* to arrival within
the flame. In the case of total consumption burners,
nonflame cells and even liquid-fuel burners, all such
physicochemical transformations must occur within
the confines of the cell, preceding the excitation/
de-excitation processes. The efficiency with which
these reactions occur for a particular metal and
matrix within the *usable* portion of the flame cell
largely controls the analytical sensitivity, and
irregular deviations give rise to the physicochemical
interference effects discussed in the following
chapter.

Flame Structure

Flames may basically be classified into two types:
diffusion or premixed.[13],[14] The choice of flame type
depends primarily upon the burning velocity of the
specific gas mixture (Tables 7.2 and 7.4). In premixed (kinetic) flames the fuel and the oxidant are,
as the term implies, thoroughly mixed prior to entry
into the combustion zone, and the burning velocity
is sufficiently slow to prevent flashback. For mixtures exhibiting a high rate of combustion, the gases
must be mixed (by turbulence) at the burning zone so
that this mutual diffusion is the slower, rate-
determining process. The latter type of flame may
be further characterized by either laminar or
turbulent gas flow and this, under ideal circumstances,
is determined by the relevant Reynolds number (Re):

$$Re = \frac{2Vr\rho}{\eta} \qquad (7.1)$$

where r is the burner radius and V, ρ and η the gas velocity, density and viscosity, respectively. Turbulent flow occurs when Re exceeds 3200.[13,15]

Figure 7.1A is a schematic of an ideal (stoichiometric) premixed, hydrogen-air flame.[13,16,17]

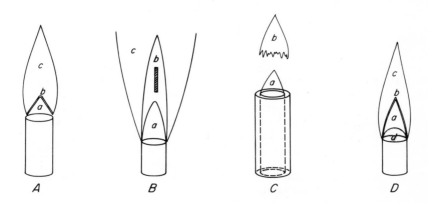

Figure 7.1.   Schematics of various premixed, laminar-flow flames.
A.   Basic hydrocarbon-air flame.
[a]Transport or preheat zone.
[b]Primary reaction zone.
[c]Post or secondary reaction zone.
B.   Fuel-rich, oxyacetylene flame.
[a]Primary (blue-cone) zone.
[b]Interconal zone showing region viewed for optimum emission data.
[c]Outer (secondary) zone.
(After Reference 18 by courtesy of the American Chemical Society.)
C.   Inert-gas separated flame.
[a]Primary zone.
[b]Secondary zone.
D.   Colloidal (liquid-fuel) flame.
[a]Transport or preheat zone.
[b]Primary reaction zone.
[c]Post or secondary reaction zone.
[d]Colloidal fuel/oxidant zone.
(After Reference 19 by courtesy of the American Chemical Society.)

Within the prereaction zone (a), the mixture is
heated to the ignition temperature prior to the
transition to the complex physicochemical primary
reaction zone (b). The latter zone is only fractions of a millimeter thick (around 10^{-1} mm for
hydrogen flames) and the temperature gradient is
of the order of 10^{5}°C/mm. The burned gas zone (c)
may constitute a secondary combustion zone for
fuel-rich (*i.e.*, nonstoichiometric) flames where
diffused atmospheric oxygen reacts with excess
carbon monoxide and hydrogen.

Basic Burner Design

     The choice of burner for flame AS analysis is
largely dictated by the desired combustion gases,
on the basis of the combustion velocity of the
mixture as noted above. If the fuel and oxidant
may be safely premixed, it is advantageous in most
cases to utilize a laminar-flow burner and to
aspirate the sample into the mixing chamber. In
this fashion, the portion which arrives at the
flame reaction zone is in a form most amenable to
efficient atomization.
     The burner dimensions and gas velocities must
be closely controlled to achieve the desired safe,
nonturbulent flame and there are special advantages
(see References 1, 15) to employing the Mekker design for AE/AF purposes. Sensitive AA trace
analysis is best performed using a premixed slot
burner (Figure 7.2A). It must be remembered that
AA analysis necessitates an approach to the ideal
homogenous absorption cell of optimum and reproducible dimensions. When the burner slit is aligned
along the source beam axis, the cell geometry will
be a compromise between the volume of flame inter-
cepted and the effective dilution of the sample
within this flame. The slot width is effectively
dictated by the combustion mixture used. The
three-slot Boling burner[20] may be found advantageous
for many applications. Increased sensitivities
noted with this design are probably due primarily
to the improved stability and chemistry of the
center flame "protected" by the flanking slots.
The surge of interest in slot type burners over the
last decade or so has prompted considerable efforts
by the manufacturers to optimize designs specifically
for AA analysis. One result of this has been the
recent trend to replace the round burner heads by
flat and recessed surface which have been shown to
improve flame stability.

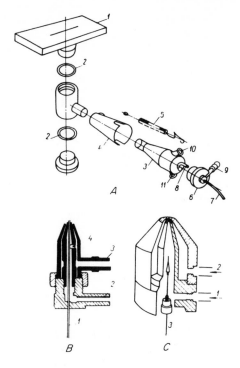

Figure 7.2. Basic AS burner designs. (Reprinted from
Reference 21 by courtesy of Iliffe Books Ltd.)
A. Premixed slot burner and nebulizer.
[1]Replaceable burner head.
[2]O-ring seal
[3]Mixing chamber.
[4]Safety shield.
[5]Retaining spring
[6]Nebulizer assembly.
[7]Aspirator capillary.
[8]Glass bead impact surface.
[9]Oxidant supply.
[10]Fuel supply.
[11]Excess sample drain.
B. Total consumption (diffusion) burner--
Beckman type.
[1]Aspirator capillary.
[2]Oxidant supply.
[3]Fuel supply.
[4]Adjusting screws.
C. High efficiency (HETCO) total consumption
burner.
[1]Oxidant supply.
[2]Fuel supply.
[3]Aspirator capillary.

Almost any oxidant-fuel mixture may be utilized in the total consumption design. The gases mix and the sample is introduced directly at the burning tip (Figures 7.2B, C). These burners are inherently safe, even when used with explosive gas mixtures, but the sample is not, as noted several times, in an ideal form for atomization within the required flame area. The point source is suitable for emission and fluorescence but less so for AA, even where several burners are arranged in series or some system of multipass optics is provided. The high velocity gases most usually employed with this type of burner produce a noisy and turbulent flame. Specialized AS applications are given below.[16,22,23]

Nebulization

The first and most important AS sample transportation step with a premix burner is that of nebulization, that is, the conversion of the sample solution to a gas-liquid aerosol. Pneumatic aspiration and nebulization are most commonly used, and many commercial designs are available which may be readily incorporated into any instrumentation package. In addition, it is a relatively simple matter to modify these components to suit specific needs, so that this is one of the few areas in AS instrumentation technique where a reasonable theoretical background will reap considerable benefits to the analyst in terms of optimization of the analysis.

It would not be appropriate here to consider the nebulization process in any great detail since this has been most adequately covered in several readily available texts.[1,15,24] Nebulization within the mixing chamber is aided by the provision of an impact surface (such as the glass bead of Figure 7.2A) which should be correctly positioned. It has been shown[24] that most of the resultant aerosol particles are less than 20μ but that some 90% of the sample volume remains as droplets in excess of this size. This latter fraction drains from the chamber unused so that, compared with a total consumption burner, only a small fraction of the liquid originally sampled reaches the flame, but it arrives in a form more amenable to free atom formation. Koirtyohann[24] has noted further that the oft quoted[25] Nukiyama and Tanasawa empirical formula for describing the nebulized droplet size is useful principally to describe the effect--in a qualitative fashion--of

the contributing physical parameters which include
air velocity, surface tension, solution density and
viscosity and the air/solution volume ratio. Changes
in any of these parameters will directly affect the
quantity of transportable aerosol and hence the
analysis signal; irregular variations give rise to
the transportation interference effects of Chapter 8.
Obviously, also, the arrangement of baffles within
the mixing chamber will affect both the droplet size
directly, and also the quantity of sample reaching
the flame by defining the pathways to be traversed.[26]
It might be expected, *a priori*, that increasing the
aspiration rate would improve the analytical signal
by increasing the rate of delivery of sample to the
cell. But the droplet size is an inverse function
of the flow rate, so that the fraction of usable
aerosol decreases with increasing air velocity.
Aspiration rates, therefore, should be optimized
for each application as illustrated in Figure 9.8.

Viscosity is the primary solution parameter
controlling nebulization, affecting the flow rate
(F) directly (r and l_c are the capillary radius and
length respectively) and also indirectly *via*
droplet-size formation:

$$F = f\left(\frac{r^4}{l_c \eta}\right) \qquad (7.2)$$

Since the viscosity coefficient (η) is related to
the temperature, heating the spray chamber or air
supply[27] is a common ploy. Minor variations in the
nebulization/feed rate, such as might be attributed
to fluctuations in the atomizer air pressure or to
partial clogging of the aspiration capillary, may
be eliminated by the use[28] of a coupled syringe pump;
but this technique cannot, *per se*, improve the
analytical sensitivity.

With recognition of some of the problems inherent in the use of pneumatic nebulizers has come
development of several alternate methods for preflame
dispersal of the sample. Of these, only ultrasonic
nebulization has been practically utilized to any
significant degree.[16,29-31] A relatively simple
model is illustrated in Figure 7.3[30] in which sample
is dropped directly onto the vibrating surface. A
70% dispersion to particles less than 1μ has been
claimed for this particular device.

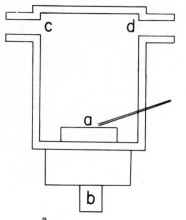

*Figure 7.3.   Nebulization chamber
of an ultrasonic
nebulizer. (After
Reference 30 by cour-
tesy of the American
Chemical Society.)*

^{a}*Vibrating surface fed by a syringe pump via a hypodermic
needle.*
^{b}*Transducer.*
^{c}*Carrier gas feed.*
^{d}*Carrier gas and sample exit.*

Desolvation and Vaporization

Sample solutions sprayed into total consumption
burners undergo desolvation, vaporization and dissociation reactions--though not necessarily to
completion--within the flame zone. Solutions
initially dispersed within premix chambers may enter
the preheat flame zone as colloidal, or possibly as
coagulated,[32] particles. In either case, the
mechanisms and reaction rates of these processes
are of major importance.

Clampitt and Hieftje[33] have presented evidence
in support of the thesis that desolvation rates are
limited by the rate at which heat can be conducted
to the droplet surface. This means that this process
is thermally controlled with a rate independent of
the initial size of the droplet so that, in this
respect, total consumption and premix burners are
probably equally efficient. An extension of this
work[34] has suggested that, under certain circumstances,
vaporization rates of the solute particles may be
subject either to a mass transfer or a thermal control. As a generalization, it would appear that the
rate of vaporization (Q) is given by:

$$Q_{RA} = 4\pi r^2 P_{RA} \left(\frac{m_{RA}}{2\pi RT}\right)^{1/2} \qquad (7.3)$$

where r and P are the radius and vapor pressure at
the flame zone temperature of a particle of the
compound RA. Compounds which boil at temperatures
below that of the ambient flame should vaporize
rapidly and efficiently, but in the reverse case
the size of the particle will be an important factor.
It should be noted that the reaction rates of these
and other physical and chemical processes of the
analyte occurring within the flame are relative to
those of the flame (for example, to the rate of
transport of the compound through the flame zone).

Interference effects on heavy metal analytes of
the common matrix anions have been extensively
documented,[32] but the relationships between the
emission/absorption signals and organic ligand
concomitants[35] are frequently ignored. This is
unfortunate in view of the prevalence of pre-analysis
concentration reactions involving chelation of the
metal ions. Koirtyohann and Pickett[36] have noted
the spatial dependence in the slot flame of enhancement effects of coexisting acid matrices. They
suggest that acids which are less volatile than
water yield droplets which laterally diffuse more
slowly than would be so for particles formed in
their absence. An alternative mechanism is referenced
below.

Some knowledge of the chemistry of flames and
of the reactions of introduced salts with the flame
radicals is of importance to anyone using flame
atomization. This topic has been reviewed in some
detail by Jenkins in a series of publications[17,37]
with particular emphasis on characteristic reactions
within the various combustion zones. Bouckaert *et
al.*[38] have recently re-emphasized the importance of
reactions with, especially, hydrogen radicals as a
control on free atomic distributions. The distribution of combustion radicals within turbulent flames
is of considerable complexity and is consequently
less well documented.[39] It is most important to
appreciate, as noted several times, that free atom
concentrations are considerably higher in certain
flame zones, depending upon the chemistry and ratios
of the combustion gases used and the physicochemical
properties of the metal species within each zone.
Oxide dissociation equilibria, for example, is a
topic which is referred to repeatedly in this chapter.
The application of separated flames is one practical
example of signal measurement from a specific flame
zone, and the flame area designated in Figure 7.1B
is another.

Oxide Dissociation and
Atomization Efficiency

It is generally conceded that dissociation
(Equation 3.35) of the diatomic oxides formed within
the flame will control the degree of atomization of
many of the elements of particular concern to the
aqueous pollutant metal analyst:

$$RO \rightleftarrows R° + O; \quad K_d = \frac{[O] \ [R°]}{[RO]} \qquad (7.4)$$

Behera and Chakrabarti[40] alone have rejected this
thesis on experimental grounds (in a fuel-rich
nitrous oxide-acetylene flame) for titanium and
several other less common metals. In many cases
the strength of the RO bond (see Table 3.3) may
approach or exceed the dissociating capability of
the flame,[41] and many of the flames commonly used
constitute oxidizing environments so that not only
will dissociation of these troublesome oxides be
inhibited but the reverse reaction of Equation 7.4
will act to diminish the supply of free atomic
species. The technology of fuel-rich, separated
and high temperature flames, and of the use of
organic solvents, has been developed primarily to
combat such specific oxide dissociation problems.
Some familiarity with these techniques is mandatory
for heavy-metal analysis and these topics are considered below. The axial enhancements observed in
the presence of concomitants in slot burners noted
above have been attributed to distortions of the
monoxide equilibria, but West *et al.*[42] consider
that the matrix delays atomization of the analyte
atoms, effectively reducing the time available for
lateral diffusion out of the viewing zone.

It has been shown in Chapter 3 that I_A (α),
I_O and I_F are definable functions of N_O but that
practical flame atomization cells are variably
efficient, so that a degree of atomization term (β)
may be defined (*e.g.*, Reference 43) as in Equation
3.37. The early attempts to determine β values for
various flames by AA have been summarized by Pickett
and Koirtyohann.[44] Both continuum and line-source
AA modes have been used, but the former have certain
theoretical advantages[45] and have been favored.
Willis,[46] for example, has noted that the commonly
utilized equations linking N_O and I_A (or α) do not
take account of hyperfine structure.

Table 7.1[45] gives some calculated β values for four flames. This type of data quantifies that which is generally well appreciated conceptually by the analyst (for example, that titanium may be determined in an acetylene-nitrous oxide flame but not in the common lower temperature flames) for the elements with which he is familiar. Such β-factor tabulations should, as the data become progressively refined, be invaluable practical aids to flame composition selection for particular analyses. But it should be noted that listings such as Table 7.1 basically document β dependence upon temperature, that is, upon the dissociation characteristics of the compounds and upon particular flame zones.

Table 7.1

Degrees of Atomization (β) Measured for Various Flames
(Data from Reference 45)

	Air		Nitrous Oxide	
Element	*Hydrogen*	*Acetylene*	*Hydrogen*	*Acetylene*
Ag	0.85	0.70	0.72	0.57
Au	0.54	0.40	0.43	0.27
Bi	0.63	0.17	0.26	0.35
Cd	0.37	0.38	0.62	0.56
Co	0.21	0.28	0.28	0.25
Cr	0.31	0.071	0.042	0.63
Cu	0.96	0.88	0.92	0.66
Fe	0.82	0.84	0.91	0.83
Ga	0.45	0.6	0.16	0.73
In	1.07	0.67	0.61	0.93
Mn	0.75	0.62	0.54	0.77
Pb	1.03	0.77	0.93	0.84
Sn	0.38	0.043	0.059	0.82
Ti	$<2 \times 10^{-3}$	$<10^{-3}$	$<3 \times 10^{-3}$	0.11
Tl	0.68	0.52	0.61	0.55
V	0.018	0.015	0.008	0.32

Some of these considerations have been elaborated by Fassel *et al.*[47] and Smyly *et al.*[48] β values are generally cited with respect to silver or copper; these metals have been shown in several studies[40,48] to be completely atomized over a wide temperature range.

Atomic Profiles in Flames

Variations of free atomic distributions as a function both of the physicochemical characteristics of the analyte and of the flame composition and structure have been noted in several instances above. Most analysts realize that some knowledge of the zones of maximum atomic concentration (*i.e.*, maximum absorption) within a given flame is an essential prerequisite for sensitive AA analysis, but, in many cases (*e.g.*, Figure 7.1B; Reference 49), the zone of optimum flame emission of heavy-metal spectra is also sharply localized. Rann and Hambly[50,51] have published atomic distribution profiles for several elements in common air-acetylene flames (see also Reference 45), and Chakrabarti and co-workers[40,52] have extended this work to include several "refractory oxide" metals (*i.e.*, those forming high dissociation energy monoxides within the flame) to illustrate the efficacy of nonstoichiometric fuel-rich flames, and have used computer processing[53] to rapidly reduce absorption and flame configuration data and to plot the required profiles.

Flame Spectra

Background flame emissions may severely interfere with AE and AF analysis lines, depending upon the type of flame and the combustion mixture employed. These molecular emissions are, in fact, a form of spectral interference as discussed in the following chapter. The least offenders are the hydrogen flames employing either air or oxygen as the oxidant which exhibit only intense OH banding at around 3100 Å (see Figure 7.4 and References 17,54). Dissociation of the fuel constituents is promoted at relatively low temperatures, and reactions of the following type serve to furnish hydroxyl free radicals within the flame reaction zone:

$$H + O_2 \rightarrow OH + O; \quad \Delta H = 17 \qquad (7.5)$$

$$O + H_2 \rightarrow OH + H; \quad \Delta H = 2 \tag{7.6}$$

This effect may be minimized by spraying aqueous solutions.[55] Use of organic solvents, on the other hand, adds considerable complexity to the emitted spectra.[56]

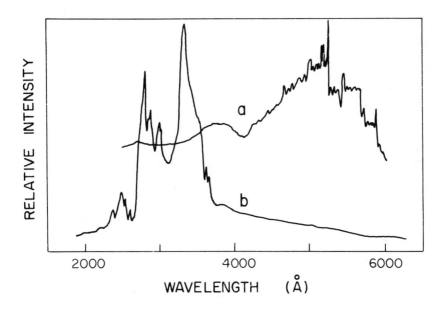

Figure 7.4. *Flame emission spectra. Relative intensity* vs. *wavelength.*

[a]*Standard oxyacetylene total consumption (Beckman) burner operated under fuel-rich mode (after Reference 18 by courtesy of the American Chemical Society).*
[b]*Argon-hydrogen entrained-air flame (after Reference 57 by courtesy of the American Chemical Society).*

The number of species present in hydrocarbon flames results in exceedingly complex background spectra from unsalted flames. However, unlike the hydrogen flames, it is usual to utilize these flames in the premix mode so that a clear differentiation between, and hence analysis from, specific flame zones is possible. Figure 7.5, for example, shows

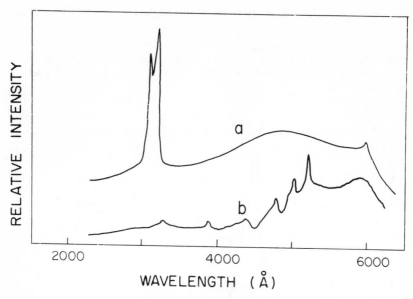

WAVELENGTH (Å)

Figure 7.5. *Flame emission spectra from interconal region of air-acetylene flame. Relative intensity vs. wavelength. (After Reference 58 by courtesy of the International Union of Pure and Applied Chemistry.)*

[a]Conventional air-acetylene flame showing OH banding and CO continuum.
[b]X15 exaggerated spectrum after flame separation (see text).

the interconal emission of a premixed air-acetylene flame, and Pungor and Cornides[54] have presented spectra illustrative of both premixed and turbulent oxyacetylene flames. The practical effect of such emissions from either of these flame types depends, however, upon the particular element of interest; for the AF determination of zinc, for example, a background-emission intensity ratio of 10:1 for premixed air-acetylene and turbulent oxyacetylene respectively has been given by Martin *et al.*[59] These problems are considerably intensified with the use of fuel-rich hydrocarbon flames, particularly where nitrous oxide is the oxidant. Since these flames are often mandatory for the AE determination of heavy metals (and commonly necessary for AA analysis also), an appreciation of these specific molecular emissions--primarily CH, OH, CN and C_2 bands-- is essential. Figure 7.4[18] shows the form of the background spectrum from a fuel-rich turbulent oxyacetylene flame. C_2 and CH bands are the most troublesome with this flame whereas CN emission predominates from nitrous oxide-acetylene mixes.[60] Potential emission problems from such high temperature flames operated in the fuel-rich mode may be circumvented to a large extent by using premixed, laminar-flow burners with judicious selection of the flame zone.

Organic Phase Samples

It is widely appreciated that enhancements of the AA analytical signals may be obtained when organic solvents are introduced to premix burners.[61-63] Since some 90% of an aqueous solution may be lost prior to arrival within the combustion zone of this type of burner, there is clearly ample scope for improvement. Increased efficiency of sample transport through the nebulization sequence may be primarily ascribed to reduced sample viscosity (Equation 7.2), but also to a more efficient production of the aerosol droplets. However, advantageous flame reactions also result, and the participating effects are undoubtedly more complex than was originally conceived.

Improvements in flame emission analytical sensitivity are even more dramatic than those realized for absorption analysis. Enhancements (see Figure 7.6) *via* the use of organic solvents by this analysis mode have long been recognized,[56,64-66] are widely applicable and frequently spectacular, and permit a precise analysis of many heavy metals which may not be practically determinable from aqueous solution. Since most flame AE analysis utilizes total consumption burners, the primary explanation of these analysis signal improvements must reside with reactions occurring within the flame proper rather than within the transportation phase as with equivalent AA enhancements, although improved evaporation characteristics undoubtedly contribute. The frequently severe cell temperature quenching associated with the introduction of aqueous samples may be totally alleviated, since the decomposition of the organic solvent adds thermal energy to the system. But the most important role of organic solvents for heavy-metal AE analysis[67,68] concerns the diminution of oxide species formed in the flame

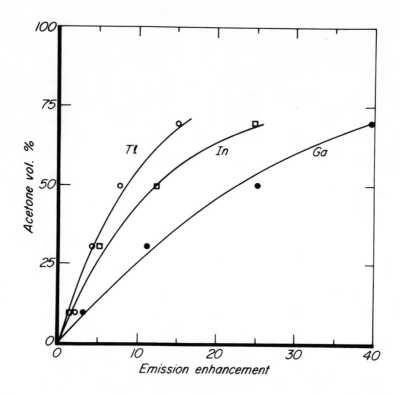

Figure 7.6. *Relationship between flame emission signal enhancement (expressed as ratio of signal from organic-aqueous mix and aqueous solvent alone) and composition of mixed acetone-aqueous solvent for Ga, In and Tl. (Data from Reference 69.)*

by mechanisms analogous to the action of the fuel-rich flames (considered below). These processes are aided inasmuch as the metal will not be initially coordinated with water or hydroxyl ligands. It should be noted however that use of organic solvents will complicate the flame background spectra[54],[64] and the choice of organic will be dictated to some extent by the relevant burning characteristics (see also Chapter 5).

Standard Air and Oxygen
Supported Flames

Flames using air or oxygen as the oxidant (Table 7.2) have long been the norm for AS flame analysis.

Table 7.2

Air/Oxygen Flame Temperatures and Velocities

Oxidant	Fuel	Inert Support	Temperature (°C)	Velocity (cm/sec)	Reference
(Entrained air)	Hydrogen	Argon	1849		70
	Hydrogen		1909		70
Air	Coal-gas		1840	55	21
	Propane		1925	82	56
	Butane		1895	83	71
	Hydrogen		2047	320 (440)[c]	56
	MAPP[a]		2200		72
	Acetylene		2300	20	21
	Cyanogen		2330	20	21
Oxygen	Hydrogen		2660	1190	73
	Coal-gas		2730		21
	Acetylene[b]	Nitrogen	2815	640	74
	Propane		2850		21
	MAPP[a]		2927	240	75
	Acetylene		3060	1130	74
	Cyanogen		4467	140	21

[a]MAPP R gas (Dow Chemical Co.): 60-70% methyl acetylene and propadiene with stabilizer.
[b]50% oxygen with 50% acetylene-nitrogen.
[c]Datum from Reference 73.

It is a prime concern in AA analysis to project the source beam through a reproducible flame zone giving maximum atomization of the test element. To this end, laminar-flow slot burners (Figure 7.2A) are invariably used and this requirement has narrowed the choice of available combustion mixtures. Air-acetylene mixes have long been standard, replacing the low-temperature flames dictated by theory and used in the early days of AA development. These latter flames (air-propane, for example) are generally

incapable of efficiently dissociating many of the
sample species analyzed in practice. A stoichiometric
mix probably yields the best general purpose air-
acetylene flame:[76]

$$2 C_2H_2 + 5 O_2 \rightarrow 4 CO_2 + 2 H_2O \qquad (7.7)$$

but fuel-rich flames are advantageous for the AA
determination of some elements, notably chromium
and molybdenum. It has been shown[77] that oxyhydrogen
flames offer few advantages for AA analysis, since
a relatively cool fuel-rich mode of operation is
required to dissociate the "refractory oxide"
species.

Oxygen flames are generally preferred for flame
AE since, as a working generalization, this method
necessitates a hotter flame environment than AA or
AF analysis. The low background emission of the
conventional hydrogen-oxygen flame is ideal for many
routine flame emission applications but is generally
insufficiently energetic for the heavy metals, and
the hotter oxyacetylene mix may be required for these
elements. Both of these flames burn at a high
velocity so that, using nonspecialized equipment,
the laminar-flow, premixed burner design is forbidden.
In addition, the oxyacetylene background spectrum is
of considerable complexity, particularly when operated
in the fuel-rich mode (Figure 7.4). In AF analysis
the low background hydrogen (air or oxygen) flames
are ideal, since high excitation temperatures are
not required. As one example, Browner *et al.*[78] have
described the AF analysis of tin using an oxyhydrogen
mix (plus an argon dilutant) in a standard Perkin-
Elmer premix burner fitted with a circular head.

Fuel-rich Flames for Emission

The AE determination of many heavy metals is
limited, not by the available flame excitation
energy,[79] but because of the formation of stable
oxides which are difficult to adequately dissociate
within the flame. It is therefore expedient to
promote the forward reaction of Equation 7.4 by
utilizing organic solvents as noted, by operating
the flame in a fuel-rich mode, or, preferably, by
both means. Through the provision of carbon species
within the flame[64] these manipulations yield a
strongly reducing environment:

$$C + RO \rightarrow R^\circ + CO \qquad (7.8)$$

The high dissociation energy of CO (11 eV; compare with $E_{D.Ox.}$ values for heavy metals given in Table 3.3) is a measure of the effective reducing ability of the free carbon species. There is, of course, a practical limit to the degree that a flame may be fuel enriched and produce the desired effects. The transition point--the critical carbon to oxygen ratio[45]--is marked by the formation of soot particles.

AE analysis of these "refractory" elements necessitates high temperature, high velocity flames (oxyhydrogen or oxyacetylene, for example). One major problem associated with the practical application of fuel-rich, turbulent oxyacetylene flames is the intense and complex background spectrum emitted [Figure 7.4(a)]. To combat this specific problem, Kniseley and co-workers have designed a premix oxyacetylene burner which allows for isolation of the optimum flame zone (Figure 7.1B[18]). The initial design of the Kniseley burner[80] is illustrated in Figure 7.7A.

Entrained-Air Flames

Hydrogen entrained-air flames, with or without the addition of an inert gas (usually argon), have been found useful for both flame emission and fluorescence work. Use of this type of flame should be considered in situations where the oxygen-hydrogen mix might be more conventionally used. Several enhancement mechanisms are operative, depending upon the analysis mode and the specific analyte. For AE analysis, improvements in the analytical signal[57,81] may be ascribed predominantly to the much-reduced background noise (Figure 7.4) noted over a considerable portion of the usable spectral range. But it should be noted that the temperatures of these flames (Table 7.2) are somewhat lower than the equivalent oxygen supported flames, so that use only with the more readily volatile metals is recommended.

Ellis and Demers[82,83] have suggested that the enhancements noted with use of an entrained-air flame (without inert support) over the oxyhydrogen flame for AF analysis (Table 7.3) are probably attributable to both the resultant larger flame diameters and to decreased collisional quenching (see Chapter 8) within the flame. In addition, it

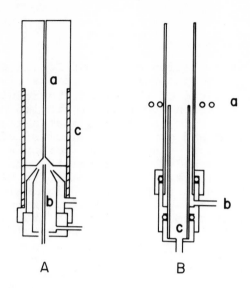

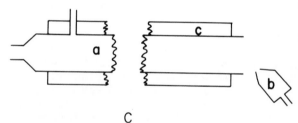

Figure 7.7.   *Some atomization devices.*
A.   *Premix, oxyacetylene (Kniseley) burner.*
(After Reference 80 by courtesy of the
American Chemical Society.)
[a]*Graphite rod.*
[b]*Beckman total consumption burner.*
[c]*Copper tube.*
B.   *An induction-coupled r.f. plasma torch for*
atomic absorption and emission.   (After
Reference 9 by courtesy of MicroMark Corp. Inc.)
[a]*Coil.*
[b]*Coolant gas.*
[c]*Sample and carrier gas.*
C.   *The absorption (Fuwa) tube.   (After Reference*
84 by courtesy of the American Chemical Society.)
[a]*Tube (c. 100 cm).*
[b]*Burner.*
[c]*Insulation.*

Table 7.3

Atomic Fluorescence Relative Detection Limit (µg/ℓ) Improvements
for Some Elements Using the Entrained-Air Flame
(Data from Reference 83)

Element	*Oxyhydrogen Flame*	*Entrained-Air/* *Hydrogen Flame*
Ag	3	1
Co	3,500	180
Cu	100	18
Fe	25,000	1,800
Mn	35,000	40
Ni	5,000	1,000
Tl	300	70
Zn	350	10

is possible that the entrained-air flame provides
more of a reducing environment, promoting the dissociation of the oxide compounds in the same fashion
as has been considered above. This latter mechanism
would aid AE analysis also and might partially ex-
plain the enhancements noted by this mode of analysis,
but the effects are unlikely to be comparable to
those obtained from fuel-rich, higher temperature
flames. Some advantages have been reported[85,86]
from the use of entrained-air (argon-hydrogen)
flames for the AA determination of those elements--
selenium, and arsenic, for example--that have major
resonance lines in the far UV region where air-
acetylene flames absorb strongly.

Nitrous Oxide-Acetylene Flames

    The search for a high temperature flame suitable
for use in premix slot burners led, in the early
sixties, to experimentation with nitrogen-diluted
oxygen flames and thence to the use of nitrous
oxide as an oxidant:[74,87]

$$5 \ N_2O \rightarrow 5 \ N_2 + 5/2 \ O_2; \ \Delta H = 102 \qquad (7.9)$$

The nitrous oxide-acetylene mix is now a *sine qua non* for the flame AA determination of the "refractory oxide" metals. The success of this flame is not, however, due primarily to the higher temperatures realized (Table 7.4). When used in even a slightly fuel-rich mode it offers an exceedingly efficient

Table 7.4

Temperatures and Velocities for Flames Utilizing
Nitrogen Oxides as Oxidants

Oxidant	Fuel	Temperature (°C)	Velocity (cm/sec)	Reference
Nitrous oxide	Propane-butane	2550	250	73
	Hydrogen	2660	390	73
	MAPP[a]	2750	140	88
	Acetylene	2880	180	76
Nitric oxide	Hydrogen	2820	30	73
	Acetylene	3080	90	74
Nitrogen dioxide	Hydrogen	1550	150	73
	Acetylene		135	73

[a]MAPP R gas (Dow Chemical Co.): 60-70% methyl acetylene and propadiene with stabilizer.

reducing environment *via* the formation of CN (and CH, NH, etc.) radicals in the flame:[89,90]

$$RO + CN \rightarrow R^\circ + CO + N \qquad (7.10)$$

This flame, in fact, achieves a carbon/oxygen ratio greater than unity with a less than 10% fuel-rich mix; an equivalent reducing oxyacetylene flame would be something approaching 50% fuel-rich. By operating an experimental burner at reduced pressures, Stephens and West[91] have demonstrated an even greater atomization efficiency for several "refractory" metals. This is added evidence for the superior reducing properties of this flame since, under these conditions, Equation 7.4 must be displaced to the right. It has also been shown[92] that the AA analytical sensitivity for molybdenum is a direct function of

the reduction of the diatomic oxide produced within
the nitrous oxide-acetylene flame. The optimum "red"
(cyanogen) flame zone is quite distinctive and easily
utilized for routine AA analysis, particularly if
flame separation techniques, as described in the
following section, are incorporated. The relatively
high temperatures produced may create problems for
some elements[93] unless precautions to minimize
ionization reactions are taken. Parsons and
McElfresh[94] have tabulated and compared a number of
experimental and calculated AA detection limits for
this flame.

The fuel-rich nitrous oxide-acetylene flame
also has proven utility for emission work,[95,96] and
Pickett and Koirtyohann[44] have further noted that
sensitivity in this mode may be enhanced by use of
the same slot burners that are used for AA analysis.
The background emission of this flame may be troublesome as is the case with all (especially fuel-rich)
hydrocarbon flames, but the intense CH bands of the
oxyacetylene flame are much reduced in the nitrous
oxide-acetylene flame and conversely for CN banding,
so that these two high temperature flames complement
each other for the AE determination of a wide range
of heavy metals.[60] As cited in Chapter 3, Winefordner[97]
has demonstrated on theoretical grounds that AE should
be a superior trace analysis technique for elements
with resonance lines greater than (approximately)
3500 Å, and it is unfortunate that this exceedingly
useful combustion mix exhibits intense band emission
in this particular spectral region.[98] Burners for
nitrous oxide-acetylene need to be sturdily constructed with the ability to dissipate the considerable heat generated. Cooled burner heads[99] are not
uncommon and designs incorporating grooves flanking
the central slot[100] provide for better entrainment
and a more stabilized flame.

Separated Flames

In the secondary reaction zone of premixed
hydrogen-air flames (Figure 7.1A), carbon monoxide
and hydrogen react with atmospheric oxygen (which
is diffused or entrained into the flame) yielding
OH and CO background emissions (Figure 7.5a[58]) which
may seriously interfere with AE and AF analysis signals derived from the interconal regions of these
flames. Separation in space of these component
zones, with the resultant improvement in the signal/
background noise ratio, provides for a considerably

more sensitive AE and AF determination of many
metals.[101]   The air-acetylene emission spectrum after
removal of the outer zone is illustrated in Figure
7.5b [note that the scale of the spectrum (b) is considerably exaggerated with respect to (a)]. The
absorption signal is not greatly affected by this
device[102] since flame emissions interfere little in
this technique. Separation of the zones may be effected by interposing some physical barrier, such as
a silica sheath (*e.g.*, References 103, 104) around
the flame but as good or better results are obtained
by employing an exterior stream of some inert gas
(Figure 7.1C) to lift the outer zone away from the
reaction cone.

Inert gas separation of standard premixed air-
acetylene flames has proven worth for both AF (*e.g.*,
Reference 105) and AE analysis (*e.g.*, Reference 106),
but the latter mode suffers in some instances from a
reduction in temperature of the reaction zone conse-
quent upon removal of the insulating exterior flame
sheath. Benefits may be realized also from separating
the nitrous oxide-acetylene flame.(*e.g.*, References
107-110). The salutary mechanism in this case, however, is less one of suppression of extraneous
interference--AE detection limits may in fact deter-
iorate in some instances because of increased CN
radiation--than of enlarging and stabilizing the "red"
zone of the fuel-rich flame so that both emission and
absorption signals of the "refractory" metals may be
enhanced.

Less Common Flames

As a generalization, flame AS analysis of heavy
metals necessitates hot flames for AE detection and,
very frequently, a reducing environment for both AE
and AA. Additionally, slow burning flames for use
in premix burners are mandatory for AA and advantageous for AE and AF. The combustion mix desiderata
follow from these requirements as discussed at length
above. There are, of course, any number of mixes
(Tables 7.2 and 7.4) which might aid a specific
analysis. The sensitivity for aluminum (a "refractory oxide" metal), for example, is some 60% greater
where nitric oxide is substituted as the oxidant in
the nitrous oxide-acetylene flame.[111]   But, in
general, the common flames described previously in
this chapter should prove adequate for most needs.

The oxycyanogen flame

$$C_2N_2 + O_2 \rightarrow 2\ CO + N_2 \qquad\qquad (7.11)$$

is exceedingly hot but is noticeably quenched with the addition of aqueous solution, and cyanogen is, in any case, both toxic and expensive. Many other "exotic" mixes[60,87,112-114] exhibit temperature and burning properties intermediate between the common flames and may occasionally be found useful.

Optimum Flames for Heavy-Metal Analysis

Figure 7.8 shows suggested combustion mixtures and modes for the determination of heavy metals under ideal conditions by AA and AF analysis. Equivalent data for flame emission are given in Figure 7.9 in two tabulations, the first (a) being the "conventional" recommendations,[15,64,115] and the second (b) the mixes and operating conditions used to obtain the optimum chemical-flame detection limit data of Table 3.6. Several trends should be noted in particular:

1. the value of "low" temperature, low background hydrogen flames (with or without entrained air) for fluorescence analysis
2. the importance and general utility of the nitrous oxide-acetylene flame for both AA and AE
3. the fact that the previously much used fuel-rich hydrocarbon flame (usually with an organic solvent) for the flame emission determination of "refractory oxide" metals is now often superseded by the premixed nitrous oxide-acetylene flame
4. the utility of separated flames for some of the "refractory oxide" metals.

Organic Liquid-Fuel Flames

Bailey and Rankin[25,116] have advocated the use of organic liquid fuels in place of gas mixtures for flame AA analysis and this procedure may commend itself to the environmental chemist since, in a few favorable cases, it could facilitate direct analysis in the field. A double nebulizer system (Figure 7.10) is required to aspirate the fuel and sample solution separately. The flame produced (using, for example, hexane as the fuel) may be termed "colloidal" (Figure 7.1D), closely approximating the premix category since the fuel colloid is

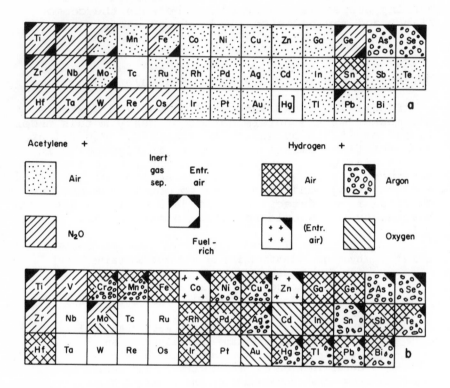

Figure 7.8　Recommended combustion mixtures.

[a]*For atomic absorption.*

[b]*For atomic fluorescence.*

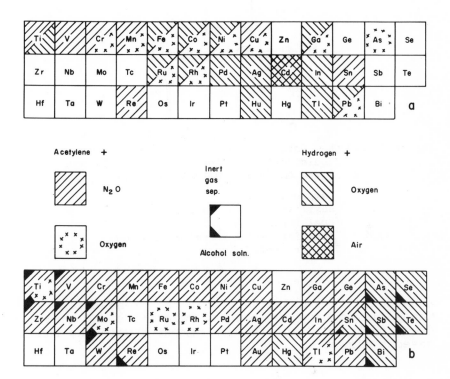

Figure 7.9   Flame emission combustion mixtures.

[a]*Standard mixes.   (Data principally from*
References 15, 64, 115.)

[b]*Mixtes utilized for the high-sensitivity data*
of Table 3.6.

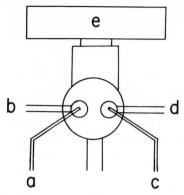

*Figure 7.10.   Dual nebulizer design for use with organic
liquid fuels.   (After Reference 25 by courtesy
of the American Chemical Society.)*
   *aSample input.
   bAuxiliary air for sample aspiration.
   cFuel input.
   dAuxiliary air for fuel aspiration.
   eBurner head.*

substantially evaporated within the preheat zone.
These flames have low burning velocities and a reduced background spectrum as compared with the common
gaseous mixes.

Absorption Tubes

   Absorption ("Fuwa") tubes were introduced[84] as
a means of increasing the residence times of the
free atom populations within the flame absorption
cell.   These devices[117-119] may therefore be considered as precursors to the more efficient nonflame
enclosed atomizers considered in the following pages.
Figure 7.7C illustrates schematically one design for
confining both flame and sample within a long narrow
tube (manufactured from Vycor or some other suitable
material) aligned along the source-beam axis.   Early
designs used conventional (*e.g.*, air-hydrogen)
point-source flames directed into the tube but ring
burners have been favored in more recent studies.[120,121]
A major disadvantage with the use of these tubes is
the complex concomitant interference effects encountered, and Rubeska[122] has suggested that these cells
constitute a marked disequilibrium environment with

regard to free radical distributions. Flame band
emissions are also notably troublesome. One suggested ingenious solution to this latter problem[123]
involves separating the flame (in this case air-
acetylene) and utilizing only the interconal portion
within the tube. Absorption tubes offer few advantages for routine AA analysis and they have, in fact,
been little used, their initial theoretical advantages
having been superseded by flameless furnace cells.

NONFLAME ATOMIZATION

For optimum sensitivity AS trace analysis, it is
necessary to generate the analyte in a free atomic
form of maximum attainable density within a cell
area defined by the optical train and detection
system of the spectrometer. To this end, flame
atomization as considered in the preliminary sections
of this chapter suffers from several severe disadvantages, especially in situations where the test
elements are present in complex matrices. First and
foremost, limitations are imposed upon the attainable
atomic vapor concentration by the continuous sample
nebulization and flame expansion and streaming processes. In the case of many of the heavy elements,
these effects may be exacerbated by the need to use
the high velocity (high temperature) combustion
mixtures (Tables 7.2 and 7.4) required to dissociate
the metal complexes present within the flames. In
the second place, the samples are introduced into
what has been aptly termed "a hostile chemical
environment."[124] Flames add foreign radicals which
participate in analytically undesirable processes
within the cell including, for example, the various
AF quenching reactions. The degree of control which
may be exercised over the chemical environment of
the flame cell is, in fact, quite limited. For
example, only small manipulations of the flame redox
environment, of the type discussed in several previous sections, are possible in practice. Kirkbright[125]
has reviewed several other practical disadvantages
inherent in the use of flames as AS atomizers.

The nonflame atomizers to be considered in the
remainder of this chapter have been designed largely
to combat the less desirable features of flame cells
that have been briefly outlined above. Before turning
to specific nonflame instrumentation, it is of interest
to consider the general characteristics of these

devices to obtain some feeling for how well and
under what conditions each approaches the ideal,
and to note caveats which may apply to individual
designs.

A few of the advocated nonflame units (such as
platinum loops and tantalum strips) atomize the
sample from a solid support into an atmosphere,
both of which are chemically inert. The majority
of commercially available atomizers, however, are
fabricated from graphite. In the former case, there
should be no disadvantageous exotic chemical reactions occurring within the cell, but this is an
a priori desideratum which depends upon the specific
metal and matrix. It has been shown in the preceding
sections that an ability to dissociate the metal
monoxides is the major factor controlling the degree
of atomization observed in flame cells, and various
techniques have been described in which the flame
chemistry may be manipulated to favor this reaction.
A primary reason for substituting flameless atomization for the conventional burners is to achieve this
objective more efficiently by supplying more thermal
energy at an accelerated rate to the sample. But
in some designs this appears to be only partially
successful. For example, the enhanced sensitivity
observed for many elements by using an additional
hydrogen flame with several filament and rod designs
clearly demonstrates the efficacy of a reducing
chemical environmental in these cases also.   Campbell
et al.[126] have, in fact, suggested that the
"appearance temperatures" of a large number of
elements correlate well with theoretical predictions
based upon reduction of the oxides by the structural
graphite (see Equation 7.8). This is an extreme
view, however, since this reaction is not thermodynamically favored in a number of cases, and the
atomization efficiency from inert tantalum and
tungsten appears to differ little from graphite over
the same temperature ranges. One important way in
which the cell graphite *does* react with some of the
analyte metals is in the formation of carbides.
This creates analytical problems only with those
elements, such as tungsten,[127] which form relatively
stable carbides, and dissociation is easily effected
for most potentially troublesome metals. It should
be noted that the above remarks are predicated on
the assumption that the reader is primarily concerned
with the determination of "refractory" heavy metals.
For AA and AF analysis, only the minimal thermal
energy necessary to atomize the sample should be

applied in order to avoid ionization problems;
filament or furnace atomization is not, for example,
a useful technique for the analysis of alkali metals.

The primary objective of generating a denser
atomic vapor within the AS reaction cell is initially
achieved *via* flameless atomization, as already noted
above, by heating the correctly positioned sample
aliquot to higher temperatures more rapidly than
is possible using conventional flame activation.
Thereafter, two options are available; the vapor
may immediately disperse or it may be physically
confined or delayed within the source beam defining
the AA/AF cell. The former mode should, *a priori*,
be notably less efficient for trace analysis than
the latter. Not only is the free atomic population
diluted more rapidly but, since the energy source
is strictly localized, the temperature gradient
away from this source is very large. Winefordner[97]
has shown that, under ideal circumstances, the
ratio (C_{cell}/C_{flame}) of *atomic* concentrations within
enclosed and flame cells for the same initial analyte
solution concentration may be represented as follows:

$$\frac{C_{cell}}{C_{flame}} = 60 \cdot \frac{V}{V_{cell}} \cdot \frac{e_{cell}}{e_{flame}} \cdot \frac{\beta_{cell}}{\beta_{flame}} \cdot \frac{F_{flame} \cdot \psi}{F_{soln}} \quad (7.12)$$

where V and V_{cell} are the volumes of discrete sample
introduced to the cell and the volume capacity of
the cell respectively, F_{flame} and F_{soln} are the flow
rates of gases and sample solution into the flame,
e represents nebulization efficiency terms, and ψ
is a flame gas expansion factor. Substitution of
reasonable practical values into Equation 7.12 yields
a C_{cell}/C_{flame} value of the order of $10^2 - 10^3$. In
practice, however, such theoretical differences are
considerably muted. For example, there is a general
need to envelope the graphite cell in an inert
atmosphere in order to protect this material from
oxidation. Enclosed cells are also invariably more
troublesome with regard to chemical matrix interference problems and spectral emissions from the
heated graphite. It should be noted that devices
in which the enclosure is the heat source are not
amenable to AF analysis because of the need to detect
the fluorescence signal at some angle to the source
beam. This stricture does not apply in cases where
the atomized sample is transported to a physically
distinct enclosing area subsequent to vaporization.

All high temperature, flameless atomization units now commonly available require that the sample solution be introduced in discrete μℓ range quantities. Atomization is effected by heating each sample aliquot individually to a high temperature very rapidly. It is frequently stated, or implied, that this ability to analyze very small amounts of sample is a very highly desirable characteristic *per se*, but, in most cases, this is mere propaganda; the mode of operation of these devices *necessitates* small volume samples, regardless of the quantity available to the analyst. For the water chemist in particular, the latter is not usually a limiting factor. Although the *absolute* detection limits presently attainable by flameless atomization are impressive, the relative solution limits that are in actuality required for water analysis are considerably less so (compare the data of Tables 7.8 and 7.9 with those of Table 7.10). Were continuous sampling, such as is used routinely with flame analysis, an available option with the more common commercially available flameless modules, better overall precision should be attainable in many practical situations. Veillon[128] has discussed the operation of an experimental tube furnace which permits continuous sampling; the Woodriff furnace (below) may also be operated in this mode.

Since the sample solution is added directly to the flameless atomization cell, the desolvation-vaporization-atomization processes discussed earlier in this chapter must be accomplished *in situ*. It is usual practice, however, to separate these reactions temporarily to some extent by thermal programming. Thus, two preatomization cycles may be employed to evaporate the sample and to remove volatile matrix constituents, respectively. In this fashion, potential band emission and particle scattering interference problems from organic matter may be ameliorated, but there is a constant danger of pre-analysis loss of volatile analyte species.[129] This is a potentially severe problem with regard to many of the pollutant metals of current major interest, lead, for example, or cadmium (particularly *via* the chloride). Very careful attention must be given to establishing the correct temperature ranges of the preatomization cycles, and these must be closely reproducible between sample batches.

Discrete sampling analysis produces a transient AS signal with a lifetime usually in the milli-second range, and this imposes severe restraints upon the detection and recording systems used. The parameters of particular relevance here are the rates of atomization, the residence time of the atomic vapor within the cell (τ_c) and the response times of the optical and electronic systems used. Theoretical treatments of this problem necessitate a partition of the common atomization devices into two groups: those in which vaporization-atomization occurs at a constant temperature, and a far larger division in which the analysis proceeds under a continuously increasing temperature regime. A relationship between τ_c and the integrated analysis signal applicable to the former group, as developed by L'vov,[2] is noted briefly in Chapter 9. However, most of the flameless modules likely to be used by the working analyst (the commercial filaments, rods and furnaces) operate under a temperature ramp, and Maessen and Posma[130] have noted that a theoretical treatment of the participating processes would be of considerable complexity in this case. It is unfortunately true that, in practice, most of the AS instrumentation in common use, having been designed specifically for continuous equilibrium-type signals, are ill equipped to handle this mode of analysis. Several theoretically superior methods of recording and processing these data have been briefly alluded to in other chapters.

A number of the readily available flameless atomization modules and some of more limited distribution, but which demonstrate potentially useful design features, are outlined in the following sections. These devices have revolutionized the practical determination of metals, especially heavy metals which are present in trace quantities in natural waters. The water chemist has now, in flameless AA and AF, an analytical tool with performance characteristics comparable to all serious rivals, as may be readily seen from a comparison of the tabulated detection limit capabilities. (It must be reiterated here that the "absolute" data are not generally directly comparable since the sample volume capacity of each cell differs.) It should be noted also that Winefordner[97] has demonstrated on theoretical grounds that many nonflame cells should offer a better performance when operated in the AF rather than the AA mode, although the former technique has not gained wide acceptance to date. The major potential problem areas common to most of these modules are as follows:

1. It is essential to add accurate sample volumes
   to the cell. Since micro additions are the rule,
   this step alone may limit the final accuracy and
   to some extent negate the intrinsic advantage of
   the discrete mode of operation. Problems associated with injecting reproducible sample aliquots
   (and particularly organic extracts from natural
   waters) into carbon filament atomizers have been
   detailed by Burrell *et al.*,[131] and Reeves *et al.*[132]
   have drawn attention to a major problem with the
   sorption of silver onto platinum surfaces of
   syringes. It should be noted also that the
   widely used nonmetal syringe dispensers tend to
   be contaminated with heavy metals such as zinc
   and cadmium.
2. The sample solution must be correctly positioned
   in the atomizer. Again, organic solutions create
   special problems: "smearing" within furnaces,
   leaking out of rod cavities, or soaking into the
   graphite cell walls.
3. Sample matrix interference problems may be
   exacerbated in the enclosed cell, and such
   effects exhibit different characteristics than
   those associated with flames. In addition most
   cells tend to exhibit memory effects.
4. The incandescent graphite may "flood" the detection system, and coil heating designs tend to
   generate spurious electronic signals.
5. As noted elsewhere, the detection-measuring
   system of most standard AS instruments is not
   well suited to handling the rapidly transient
   signals produced by the majority of these
   atomization devices.

Boat and Cup Reservoirs

The Perkin-Elmer Corporation tantalum boat
design[133] is one of the first commercially available
discrete sample atomizers still in widespread use.
This device utilizes a conventional slot burner to
supply the necessary atomization energy and is
capable of accommodating a relatively large sample
aliquot. In use, around 1 mℓ or so of solution is
pipetted into the metal boat. The sample is then
evaporated in the outer fringes of the flame and
finally atomized in the flame centered in the
incident beam. Only metals atomized at relatively
low temperatures may be determined by this procedure,
however, since the limited thermal output of the
flame is partially dissipated by the boat and the

maximum absorption signal is a function of the
length of time the heat is applied. Fortunately,
several environmentally important metals fall in
this category and good results have been reported
for silver, lead and zinc.[134],[135] Figure 7.11
illustrates x-t recorder traces for the latter
element. Although the mode of operation continues

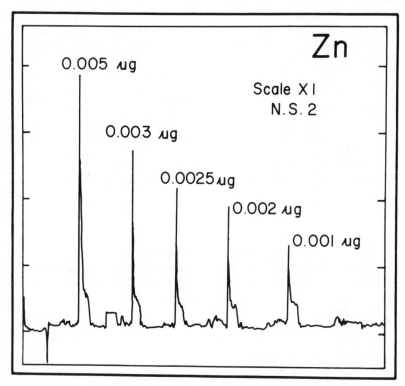

Figure 7.11. *Standard zinc % absorption data using the*
Perkin-Elmer Corporation tantalum "sampling
boat" atomization device.

to rely upon a flame to promote atomization, the
important feature is an enhanced concentration of
sample in the theoretical absorption cell compared
with that possible by the best continuous flow
system. It is possible to some extent to increase
the detection limit capability by simply increasing
the quantity of evaporated sample in the boat. As

a simple example of a discrete atomizer, the utility
of the flame-activated boat reservoir is strictly
limited, however. The low activation temperature
enhances the importance of matrix interferences[136]
and the commercial Perkin-Elmer module incorporates
some poor design features which make it difficult,
for example, to maintain the alignment of the boat
along the optic axis during analysis or to reproduce
the cell geometry for subsequent analyses. In spite
of these objectives, many routine water analyses
may be performed more rapidly and efficiently with
this device than with conventional flames. Table
7.5 lists published "boat" detection limits for
several elements but it should be noted that little
practical work has been reported for metals other
than the three cited above.

The "Delves' cup" atomization system,[137] which
has been used most particularly for the analysis of
lead in biological and environmental solutions,
utilizes flame-activated atomization from a circular
nickel reservoir much as in the boat sampler above.
In this case, however, the vapor passes into a
confining, open-ended tube aligned along the source

Table 7.5

Atomic Absorption Detection Limits:
Discrete Sampling Mode Using Unenclosed Boat Atomizers

Element	Line (Å)	Detection Limit pg[a]	Detection Limit μg/ℓ[b]	Reference
Ag	3281	400	2.0	138
As	1937	20,000	20.0	133
Bi	2231	3,000	3.0	133
Cd	2288	5	0.05	139
Hg	2537	20,000	20.0	133
Pb	2833	100	1.0	139
Tl	2768	1,000	10.0	139
Zn	2139	5	0.05	139

[a]Absolute detection limits (in picograms) under the given
conditions.
[b]Aqueous solution detection limits.

beam of the AA spectrometer. This unit, illustrated in Figure 10.7, is intermediate in some respects between an "open" atomization system and the enclosed tubes and furnaces described in the following sections.

The cup reservoir option available with the Varian Techtron Model 63 rod atomizer (considered below) probably delivers close to the ultimate AA sensitivity performance possible with this type of *open*, flameless cell. Use of this device (illustrated in Figure 7.18B) permits atomization of sample volumes of up to 20 µℓ, but any potential gain in sensitivity here must be balanced against the ability of the spectrometer to process the analytical signal. It might also be possible to digest solid sample, biological tissue, for example, directly in the cup prior to analysis. The reservoir design of Figure 7.18B is not suitable for AF analysis but could be easily modified.

Unconfined Filaments and Loops

A considerably more elegant and efficient open, nonflame reservoir was introduced in 1969 by West and Williams.[140] This device consists of a small graphite filament some 2 mm in diameter and 20 mm in length rigidly supported between water-cooled electrodes and directly heated electrically so that temperatures in excess of 2500°C (see Figure 7.12) may be attained very rapidly. Small sample volumes

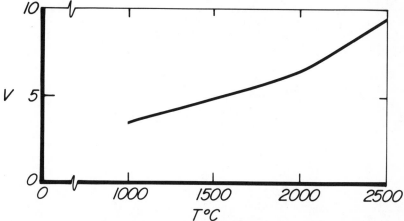

Figure 7.12. The carbon filament atomizer; filament temperature
(T°C) vs applied voltage (V). (Data from Reference 141 by courtesy of Elsevier Publishing Company.)

(around 1 μℓ) are introduced into a slot depression
on the upper surface and, since the filament is
maintained *in situ* within the source beam for
either AA or AF mode analysis, the cell geometry
is permanently fixed. It was found necessary to
envelope the reservoir in use with an inert gas
atmosphere to prevent oxidation of the filament
material. The original module[140,141] was housed
within a glass chamber which was purged immediately
prior to use with either argon or nitrogen. This
design is illustrated in Figure 7.13A.

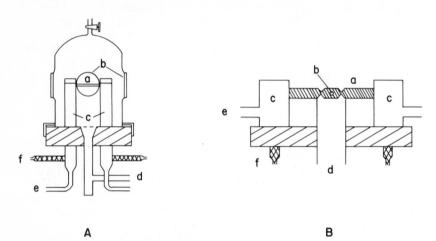

A   B

Figure 7.13.   *The carbon filament atomizer.*
   A.   *Original West module with physically con-
        fined inert atmosphere. (After Reference
        141 by courtesy of Elsevier Publishing
        Company.)*

        [a]Carbon filament
        [b]Silica window
        [c]Electrodes
        [d]Inert gas purge
        [e]Power supply connections

   B.   *"Mini-Massmann" modification of carbon
        filament with unconfined inert gas stream.
        (After References 142, 143.)*

        [a]Carbon filament
        [b]Sample cavity
        [c]Electrodes
        [d]Inert gas supply
        [e]Coolant
        [f]Power supply connections

It was quickly discovered[144] that detection limits could be improved by a factor of approximately two by eliminating the glass dome and, instead, sheathing the filament with a continuous, unconfined laminar flow of the inert gas. Either argon or nitrogen appear to be equally efficient for this purpose. At the same time, experiments conducted to determine the effect of organic matrices on the analyte absorption signals[145] indicated that programming of the thermal cycle, and particularly the addition of a preatomization, low temperature "ashing" cycle, was considerably advantageous for most practical samples. These improvements subsequently formed the basis of the Varian Techtron Model 61 atomizer which can accommodate either the open West-type filaments or enclosed furnace-type designs (Figure 7.13B).

Very careful attention must be paid to optimizing the dry, ash and atomization cycles for each specific element and matrix.[146] Figure 7.14 is an arbitrary example of the dependence of zinc absorption on the applied atomization temperature. Flameless atomization techniques have been developed, as noted previously, primarily for application to species which are moderately difficult to atomize by flame procedures. The maximum atomization temperatures attainable by most filaments, however, are somewhat less than those of conventional furnaces and, although one recent comparative study[147] has suggested that molybdenum may be more efficiently atomized than zinc, most use to date has been for the analysis of the more volatile heavy metals such as silver, cadmium, lead and zinc. In these cases, extra care must be taken to ensure that the metal is not unwittingly lost during the preatomization thermal cycles (cadmium as $CdCl_2$ is a notorious example[148]). Reeves *et al.*[149] have demonstrated that silver and copper may be sequentially atomized to yield distinct absorption peaks. Multielement analysis of this type may well be feasible for "simple" solutions, but it should not be attempted with natural water samples, especially for trace constituents in concentrated matrices. This topic is considered briefly below. It should be noted also that the electrical conducting properties of the graphite filaments change with use so that these should be frequently replaced or the atomization cycle should be reoptimized at frequent intervals.

It was suggested above that the ability (in fact, necessity) of atomizing $\mu\ell$ range samples is not a particularly useful characteristic for most

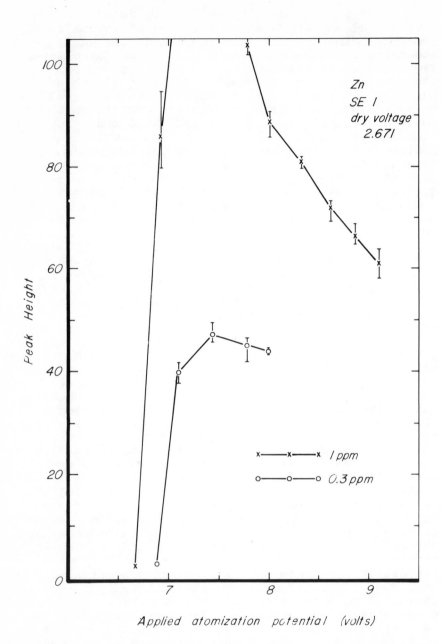

Figure 7.14. Recorder peak height <u>vs</u> applied potential for 1 μℓ quantity zinc standards atomized from a West-type filament.

analysts concerned with natural waters. However,
if a prior concentration is required, or if the trace
metals of interest must be separated from some
troublesome matrix, then this feature permits such
manipulations to be performed on small volumes of
primary sample with the attendant advantages outlined
in Chapter 4. Figure 7.15[150] illustrates x-t traces
for standard additions of zinc to seawater which
have been subsequently extracted and concentrated
by a factor of 10 into chloroform. Organic matrices
do not pose any particular problems[145] so long as
the necessary thermal cycle programs are carefully
researched. It is possible[151] to modify the filament
to increase the cavity capacity by about an order of
magnitude and hence better the relative solution
detection limits. An increase in the reservoir size
much beyond this, however, negates many of the
original advantages of using an open-slot configuration. If, for any reason, it is necessary to atomize
larger sample aliquots for AA analysis, enclosed-type
atomizers should be investigated. Several other
variations of the basic West design have been advocated, usually with the primary objective of effecting
a more rapid temperature rise *via* reduced resistance
of the filament. Molner *et al.*,[152] for example, have
experimented with various grades of graphite and have
reduced the thickness of the filament in the vicinity
of the reservoir, and Cantle and West[153] have constructed an all-tungsten element. Use of glassy
carbon should probably produce similar advantages.

The filament atomizers incorporate no mechanisms
for physically confining the atomic vapor, so that
rapid dispersal occurs, and this is aided by the
flushing action of the inert gas stream. Moreover,
and in contrast with the dynamic flame-activated
cells, no further thermal energy is supplied to the
vapor once it is removed from the immediate vicinity
of the filament, so that rapid condensation of the
analyte atoms with like or concomitant species is
promoted. With certain matrices, this latter
phenomenon gives rise to considerably complex
inhibitory interference effects, as documented by
Alger *et al.*[154] Since such interferences are specifically associated with the vapor phase, it should
be noted that the more marked depressive effects are
(again in contrast with flame cells) associated with
the readily volatilized matrix elements. Recognizing
these problems, Anderson *et al.*[155] have recommended
viewing and measuring the radiation immediately
above the filament surface (within about 0.5 mm).
Winefordner and co-workers[132,156] have shown that this

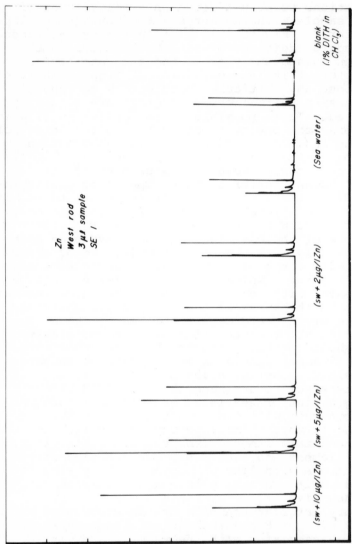

Figure 7.15. x-t recorder trace for standard additions of zinc to seawater, extracted into chloroform (with a X10 concentration), and atomized from a West-type filament. The second peak of each doublet only is a datum, the first results from interception of the source beam by the sampling syringe (Reference 150).

approximately exponential decrease of the free atomic
population above the filament can be arrested to some
extent by enveloping the reservoir in an argon-
hydrogen entrained-air flame. This procedure was
initially introduced by Amos *et al.*[157] *via* the simple
expedient of adding hydrogen to the inert argon stream
and allowing the mix to ignite spontaneously during
the atomization cycle. The attenuation of the free
atom population decay by this means may be attributable to two major mechanisms. Monoxide formation
is discouraged by excluding oxygen and dissociation
is aided by reduction reactions of the following
type in a fashion analogous to that previously
described for flames:

$$RO + H_2 \rightarrow R^\circ + H_2O \qquad (7.13)$$

It has also been noted[147] that absorption interference
signals from volatilized filament carbon are absent
when an auxiliary hydrogen flame is employed, and
this is probably due to removal of the particulate
carbon as gaseous hydrocarbon products. Hydrogen
diffusion flames can also be used in conjunction
with the furnace-type filaments described below but
the gain in analytical sensitivity is, as might be
expected, considerably less in these cases.

Since the atomic vapor is irradiated and viewed
above the filament, these atomizers are also very
suitable for AF mode analysis (see Reference 158
and Table 7.6). The furnace AF detection limit
values presented in the latter table were obtained
using a vertical "open" cell arrangement as illustrated in Figure 7.21B, so that these data are
reasonably comparable. It should be noted that
quenching interferences will be reduced if a
monatomic purge gas is used in place of nitrogen.
Several applications have also been given[97,103]
in which sample solutions have been atomized for AF
analysis from simple heated loops constructed from
some inert support material, usually platinum. This
procedure is equally applicable to AA analysis and
might offer a less complex alternate to the enclosed strip devices (below) for some routine water
analysis problems. Aldous *et al.*[162] have used such
loop samplers for induction plasma analysis; these
detection limit data are included in Table 7.7.

Table 7.6

Atomic Fluorescence Absolute Detection Limits (pg):
Discrete Sampling Mode

Element	Line (Å)	Massmann Furnace[a]	Carbon Filament[b]	Element	Line (Å)	Massmann Furnace[a]	Carbon Filament[b]
Ag	3281	1.5	1	Ga	4172		50
As	1937		500[c]	Pb	2833	35[d]	
Au	2676		10[c]		4058		10
Bi	3068		10	Sb	2311	200	
Cd	2288	0.25	0.15	Tl	2768	2,000[d]	
Cu	3248	450	5[c]		3776		50
Fe	2483	3,000	2[c]	Zn	2139	0.04	0.02

[a] Data from Reference 159; precision as defined by Reference 160 at 99.7% confidence level.
[b] Data from References 141, 144, 2x background; West-type filament.
[c] Analytical line not defined in reference; Reference 157.
[d] Lines not commonly utilized for flame fluorescence.

Table 7.7

Atomic Emission Absolute Detection Limits (pg):
Discrete Sampling Mode[a]

Element	Line (Å)	Morrison Furnace[b]	Sample Loop[c]	Element	Line (Å)	Morrison Furnace[b]	Sample Loop[c]
Ag	3281	100		Ga	4033	1,000	
As	2350		42	Hg	2537	10	16
Au	2676	1,000		In	4102	1,000	
Bi	3068	500[d]		Pb	2614		120[d]
Cd	2288		0.2		3640	1,000[d]	
	3261	10		Sb	2528		500
Cr	4254	1,000		Te	2143	1,000	
Cu	2179		1,000[d]		2383	1,000	
Fe	2483		300	Zn	2139		80
	3720	10,000			3076	10[d]	

[a] Techniques comparable with flame emission.
[b] r.f. furnace; data from Reference 161.
[c] Data from Reference 162; Winefordner-type (Reference 163) Pt sample loop; induction plasma; 0.12 µℓ sample.
[d] Lines not commonly utilized for flame emission.

Ambient and Low Temperature Tubes

In Chapter 10, analytical methods are described for arsenic, mercury and selenium in which the trace-metal species present in natural waters or in associated solid phases may be introduced into the AA atomization cell as reduced gaseous products. This is an entirely different type of nonflame (or, at least, nonburner) analysis, but one which, because of the current interest in these pollutant metals, has gained rapidly in popularity.

For mercury analysis, the sample is chemically treated to release metallic mercury which is sub- sequently flushed into a glass tube containing silica end-windows aligned along the source beam. No additional atomization energy is required, and the tube serves merely to confine the vapor for a period while the absorption signal is measured. It may, however, be necessary to concentrate the mercury prior to introduction into the tube, and some methods for accomplishing this are noted in Chapters 5 and 10. Two variations of this basic procedure are in common use. The vapor may be recirculated through the cell to yield an equilib- rium signal (Figure 10.3) akin to that obtained for continuous flame sampling or a discrete "burst" technique may be used.[164] Practical problems associated with water vapor droplets entering the tube and pre-analysis sorption losses of the mercury are cited in Chapter 10.

Arsenic and selenium are common examples of metals which yield volatile hydrides on reduction (see Chapter 5) which may be readily dissociated in low temperature Vycor or quartz tubes for absorption analysis. The gaseous hydrides are swept into the open-ended tubes as discrete samples using an inert gas. Goulden[165] has noted that dissociation is accomplished more efficiently if air is added to the purge (*i.e.*, an argon diluted air-hydrogen combustion, utilizing the hydrogen generated during the reduction step), than is the case if air must diffuse into the tube.[166] The analyte hydrides may also be concentrated prior to analysis (Chapter 5).

Furnace-Type Filaments and Rods

Amos and co-workers[157,167] introduced a mini- furnace atomizer which they prepared by drilling a small (1.5 mm) transverse hole through the West filament. This filament design (generally termed

Table 7.8

Atomic Absorption Absolute Detection Limits (pg):
Discrete Sampling Mode Using Enclosed Carbon Filament
(Data from References 157, 142, 168)

Element	Line[a] (Å)	Limit[b] (pg)	Element	Line[a] (Å)	Limit[b] (pg)
Ag	3281	0.2	Mo	3133	40
As	1937	100	Ni	2320	10
Au	2428	10	Pb	2833	5
Bi	3068[c]	7	Pd	2476	200
Cd	2288	0.1	Pt	2659	200
Co	2407	6	Sb	2311[d]	30
Cr	3579	5	Se	1960	100
Cu	3248	7	Sn	2246	60
Fe	2483	3	Tl	2768	3
Ga	2874	20	V	3184	100
Hg	2537	100	Zn	2139	0.08
Mn	2795	0.5			

[a]Inferred analysis line; data not given in references.

[b]Detection limit not defined.

[c]This line is not utilized for flame absorption.

[d]Minor flame absorption line.

a "mini-Massmann") combines to some degree the advantages of filament atomization with physical
confinement of the vapor. It is optionally available with the Varian Techtron Model 61 atomizing
unit (Figure 7.13B; References 142, 143). By
comparison with the West-type filament, analytical
sensitivity is somewhat improved, but precision is
poorer[131] and matrix interference effects tend to
be exacerbated.[169] In general, little advantage
accrues to use of the hydrogen diffusion flame
described above. The unit is operated in a somewhat
similar fashion to that employed for the West filament, including continuous inert gas streaming
around the reservoir (see Figure 7.13B). This flow
tends, of course, to sweep atomic vapor out of the
cell, as illustrated by Figure 7.16, and must be
closely regulated. Recorder traces for a series of
aqueous cadmium standards are shown in Figure 7.17,
and a suite of absolute detection limit data is
listed in Table 7.8.

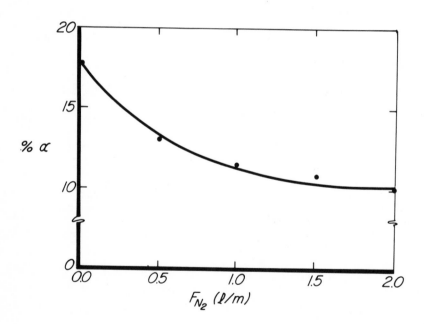

Figure 7.16. *Varian Techtron Model 61 carbon filament atomizer,*
mini-Massmann mode. Inert gas flow rate
(nitrogen, ℓ/min) vs absorption for 50 μg/ℓ
cadmium solution.

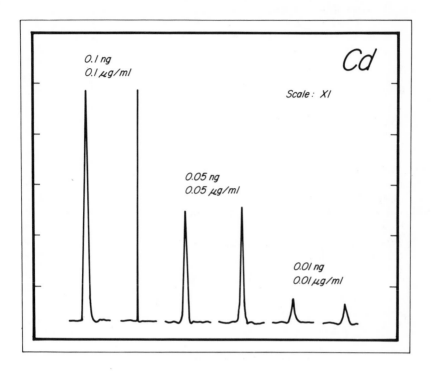

Figure 7.17. Standard cadmium % absorption data using the Varian Techtron Model 61 atomizing unit in the mini-Massmann mode.

The specific practical disadvantages of the mini-Massmann atomizer are as follows:

1. Only very small sample volumes may be accommodated (around 1 µℓ) and analytical precision is dependent upon the analyst's ability to reproduce identical aliquots in the reservoir. The design of the sampling port forbids the use of all commonly available syringes.
2. These problems are compounded when organic solvents are used. Only glass (Drummond-type) pipettes are suitable for transferring such samples to the filament[131] and these solutions tend either to run out of the cavity or to soak into the graphite making reproducible atomization an impossibility. To combat the latter

problem, prior impregnation of the rod with
xylene has been advocated[167] but better
results may be obtained by adding the solvent
at some set time during the drying cycle so
that the solvent is evaporated on contact.
The temperature must not be too high, however,
else spattering of the sample will occur.
3. Since it is not easily possible to shield the
heated element, "flooding" of the detection
system may result if high atomization temperatures are required.

An improved version of this type of atomizer
has now been placed on the market by Varian Techtron
(the Model 63 carbon rod atomizer; Figure 7.18).
The reservoir of this unit has a higher volume
capacity (around 5 μℓ), and use of a pyrolytic

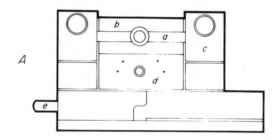

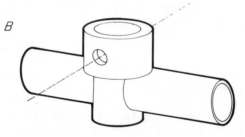

Figure 7.18. Varian Techtron Model 63 carbon rod atomizer.
A. Work-head assembly.
*   aCarbon tube-type cell*
*   bLight shield*
*   cElectrical terminals*
*   dInert gas or hydrogen inlet and diffusing*
*   screen*
*   eCooling water inlet*
B. Higher sample capacity (to 20 μℓ) carbon
*   cup-type cell.*

graphite prevents both soaking of sample and diffusion of atomic vapor into the cell walls. A light
shield is also incorporated to block extraneous
radiation. Relative solution detection limits
given by Parker[179] are included in Table 7.9.

As in the case of the West filament, the carbon
rod atomizer has been chiefly used to date to analyze
those heavy metals which may be atomized at relatively
low temperatures. A maximum attainable temperature
of something less than 3000°C results in incomplete
atomization of the "refractory" metals so that the
inter-sample operating parameters must be reproduced
in an exceedingly careful fashion. Memory effects
and the formation of carbides of, for example,
molybdenum and vanadium,[180] may create additional
practical problems. For this group of heavy metals,
use of the auxiliary reducing hydrogen flame may
improve analytical sensitivity,[181] and Morrow and
McElhaney[182] have noted that the addition of methane
to the purge gas also helps, presumably *via* a
similar mechanism. Obviously, a basic modification
of the existing equipment to generate higher atomization temperatures would[182] *per se* avoid or alleviate
most of these problems. Several analysts have noted
that AA sensitivity is enhanced by using an argon
purge in place of nitrogen. This has also been
previously recommended for filament AF analysis but
the salutary mechanisms must differ in each case.
Background absorption effects are considerably less
troublesome with the rod than with conventional
furnace atomization (below); this is a convenient
situation for users of single-beam equipment since
automatic continuum background correction is not
possible. Matrix interference problems[181] are likely
to be enhanced where the concomitant atomizes over
the same temperature range as the analyte. The
caveat concerning losses during the nonatomization
thermal cycles, previously emphasized with regard
to the open filament, applies equally in this case,
and the analyst is urged to consider a chemical
separation (Chapter 5) of the test metals from
complex or concentrated matrices prior to
introduction into the cell.

Enclosed Boat and Strip Reservoirs

Donega and Burgess[178] have designed a discrete
sample, nonflame atomization module incorporating a
physical confinement of the atomic vapor. 50-100 μℓ

Table 7.9

*Atomic Absorption Relative Detection Limits ($\mu g/\ell$):
Discrete Sampling Mode*

Element	Line (Å)	Woodriff Furnace[a]	Welz/PE Furnace[b]	Carbon Rod[c]
Ag	3281	0.2	0.003	0.04
As	1937		30[f]	20
Au	2428		0.08	2
Bi	3068[d]		0.1	1
Cd	2288	0.3	0.001	0.02
Co	2407		0.04	1
Cr	3579		0.05	1
Cu	3248	6	0.01	1
Fe	2483	6	0.03	0.6
Ga	2874			0.4
Hg	2537			20
Mn	2795	6	0.01	0.1
Mo	3133			8
Ni	2320	1	0.1	2
Pb	2833	1.1	0.06	1
Pd	2476			40
Pt	2659			40
Sb	2311[e]		0.2	6
Se	1960			20
Sn	2246		25[f]	10
Tl	2768		0.14	0.6
V	3184			20
Zn	2139	0.8	0.0006	0.02

[a] Calculated from data given in Reference 173.

[b] Data from Reference 177 except as noted; 100 $\mu\ell$ sample.

[c] Inferred analysis lines; data from Reference 179, 5 $\mu\ell$ sample.

[d] Line not utilized for flame absorption.

[e] Minor flame absorption line.

[f] Data from References 174, 175.

samples are added to either tantalum or graphite boat reservoirs which are electrically heated in the same way as the previously described graphite filaments and rods. The reservoir is contained within a quartz tube absorption cell purged with an inert gas maintained at sub-atmospheric pressures. Detection limit data for some heavy metals using this device have been included in Table 7.10.

Table 7.10

Atomic Absorption Absolute Detection Limits (pg): Discrete Sampling Mode Using Various Furnace Devices[a]

Element	Line (Å)	L'vov Crucible[b]	Massmann Furnace[c]	Woodriff Furnace[d]	Welz/PE Furnace[e]	Donega Tube[f]
Ag	3281	0.03	0.8	12	0.25	
As	1890		1000			
	1937				1000	
Au	2428	70			8	
Bi	3068[g]	25	200		10	
Cd	2288	0.003	2	13	0.1	
Co	2407	7.5			4	
Cr	3579	50			5	
	3594					6
Cu	3248	6.3	10	320	1	3
Fe	2483	25	20	280	3	
Ga	2874	25				
Hg	2537	450	200			
In	3039	8	200			
Mn	2795	0.02	8	270	1	3
Mo	3133	50				3,000

Table 7.10, continued

Element	Line (Å)	L'vov Crucible[b]	Massmann Furnace[c]	Woodriff Furnace[d]	Welz/PE Furnace[e]	Donega Tube[f]
Ni	2320	25		73	10	
	3415					3,000
Pb	2833	30	10	57	6	
Pd	2476	50				
Pt	2659	250				300,000
Rh	3435	63				
Sb	2311[h]	50	100		8	
Se	1961	200	2000		1500	
Sn	2246					
	2863	10				
Te	2143	7.6				
Ti	3653	500				
Tl	2768	2.5	40		14	
V	3184					600
Zn	2139	1.0	0.8	39	0.06	

[a] Detection limits in picograms as given in cited references; these data columns are not directly comparable.

[b] Data from References 170, 171; precision and volume of solution used not defined.

[c] Data from References 159, 172; precision as defined by Reference 160 at 99.7% confidence level.

[d] Data from Reference 173; precision as for (c) at 95% confidence level; 50 µℓ sample utilized.

[e] Introduced by Welz, marketed by Perkin-Elmer Corporation as modules #HGA-70 and -2000. Data from References 174-177, detection limit defined as 2x standard deviation of background; 100 µℓ sample utilized.

[f] Data from Reference 178; quartz tube with controlled atmosphere and various boat materials; detection limit not defined.

[g] Line not utilized for flame absorption.

[h] Minor flame absorption line.

A modification of this unit is now offered by Instrumentation Laboratory Inc. A tantalum strip with a reservoir depression is used and the vacuum system has been eliminated (Figure 10.8; References 183-185). It is questionable whether the tube enclosure is of marked benefit here since the heat source does not surround the vapor as is the case with the furnaces described in the following section. The negative temperature gradient away from the strip must be comparable with that observed for the West filaments so that rapid diminution of the atomic populations by condensation reactions will occur as previously described. In addition the vapor is flushed from the cell by the inert gas stream, but this applies equally to all the described flameless atomizers with the exception of the closed cycle, cold vapor mercury cells. Use of tantalum as the reservoir construction material limits the maximum attainable temperature to around 2500°C, so that this method would appear, *a priori*, to be better suited to the determination of the more readily atomized pollutant metals such as lead.[184] However, several applications to more "refractory" elements have also been described (aluminum,[186] for example, and chromium[187]).

Furnaces

The first practical procedure which employed both flameless atomization and a physical confinement of the atomic vapor for AA analysis was developed by L'vov, based upon the earlier King furnace.[188] This work appeared in the English-language spectroscopy literature in 1961[189] but was thereafter virtually ignored for a number of years. An increasing need for methods suitable for the analysis of metals at very low concentration levels provided the impetus for research into techniques which, like the L'vov crucible technique, circumvented some of the problems inherent in the use of flame cells. A later version of the basic L'vov furnace,[2,170] in which the cell is lined with pyrographite to prevent diffusion of the vapor into the walls, is illustrated in Figure 7.19 (the earliest model utilized an internal metallic foil lining). It should be noted that sample loss in this fashion is a constant problem to be guarded against in all nonflame carbon cells. The crucible of Figure 7.19 is approximately 30-50 mm in length

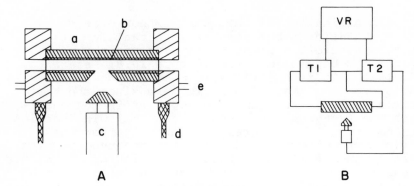

*Figure 7.19.   The L'vov furnace (after Reference 170 by
courtesy of MicroMark Corp. Inc.).*
A.   The atomization unit
   aGraphite crucible
   bCrucible lining
   cSample electrode
   dElectrical connections
   eCoolant
B.   Electrical connections
   VR - voltage regulator
   T1, 2 - Step-down transformer

with an internal diameter of 2.5-5.0 mm. The sample
is placed on the auxiliary carbon electrode (which
is also chemically treated to prevent absorption of
the solution) and dried. The tube is heated and,
simultaneously, the sample electrode is fitted into
the tube and independently heated to vaporize the
sample. This unique heating arrangement is designed
to encourage very rapid sample vaporization (in
around 0.1 sec) and L'vov[2],[171] has paid considerable
attention to the problems of accurately and efficiently measuring the resultant rapid pulse signals.
This work is referenced again in Chapter 9. Detection
limits[170] are included in Table 7.10 and claimed
sensitivity values are listed in Table 9.2; the
latter data do not appear to be readily comparable
with that obtained from other furnace types. The
pioneering work of L'vov included the use of a continuum reference source with associated circuitry
to permit simultaneous correction of nonresonance
background absorptions and this feature is now
usually included in dual-beam commercial
instrumentation.
    The L'vov crucible technique necessitated the
atomization of only very small, discrete samples

and all succeeding instrumentation produced in
commercial quantities has utilized this sampling
mode also, so that it is often forgotten that continuous sampling may be, in some cases, the analytically superior procedure. In this respect, the
Woodriff furnace shown in Figure 7.20B[190] (see also
improved design[191]) is of considerable interest
since, while discrete sample aliquots may be introduced in the usual manner,[173] solutions may also be
nebulized into the argon purge stream[190,192] in a
fashion analogous to flame cells. Woodriff also

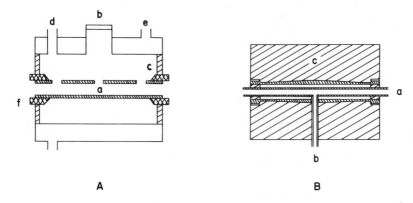

A   B

Figure 7.20.   *Schematics of commercially available AA*
atomization furnaces.
   A. *The Perkin-Elmer Model HGA-70 and -2000; by*
      courtesy of Perkin-Elmer Corporation.
      ªGraphite absorption tube
      ᵇSample port
      ᶜInsulation
      ᵈInert gas purge
      ᵉCoolant
      ᶠElectrical connections
   B. *The Woodriff furnace (after Reference 190*
      by courtesy of the MicroMark Corp. Inc.).
      ªGraphite absorption tube
      ᵇSample port
      ᶜInsulation

early appreciated the advantages of using a continuum
reference beam.[193] This furnace has not, unfortunately,
been widely evaluated, and it is too bulky (20 x 30
cm) to fit easily on most standard AA instruments.
The designers' absolute and relative detection limit
and sensitivity data are given in Tables 7.9, 7.10

and 9.2 respectively. Temperatures up to 3000°C
may be maintained (although lower working temperatures are adequate for most elements) and the furnace
has been designed to resist temperature fluctuations.
This is important since samples are inserted or
nebulized into a *preheated* absorption cell and
atomization does not proceed on a rising temperature
ramp as is the case with all other flameless atomizers
considered in this chapter.

Robinson *et al.*[194] have attempted to extend the
utility of an atomizer, designed more particularly
for air analysis,[195] to include natural water samples.
From the supporting published data, it would appear,
however, that the conversion has been only partially
successful. This flameless atomization unit uses
an r.f. coil to heat a carbon bed located in a side-
arm of the absorption cell tube, and it has been
suggested[196] that this design permits atomization
of metals from an organic matrix without resorting
to a preliminary "ashing" cycle. A maximum attainable temperature of less than 1500°C severely limits
the potential worth of this equipment for aqueous
heavy-metal analysis. Morrison and Talmi[161] have
also used an r.f. furnace to atomize samples for AA
analysis, but have additionally obtained emission
data for many elements (Table 7.7) by utilizing the
r.f. field to generate an argon plasma excitation
source.

The heat source of the Massmann[159,197] furnace,
illustrated in Figure 7.21A with the graphite tube
(55 x 1.5 mm) in the AA analysis position, surrounds
the sample in the conventional fashion. Samples are
introduced through the side port, and the cell
temperature may be raised very rapidly to around
2500°C maximum. A dynamic inert gas atmosphere is
maintained as with the other atomizers described.
The AA detection limit data given in the original
publications are reproduced in Table 7.9.

The Massmann method, adapted by Welz and
Wiedeking,[174] evolved into the commercially available
Perkin-Elmer Corporation Model HGA-70 and -2000
atomizers (Figure 7.20A; Reference 198), probably
the most commonly used flameless atomizers at the
present time. Tables 7.9, 7.10 and 9.2 list early
absolute and relative detection limit and sensitivity
(slope of the common form of the working curve)
data.[174-177] The graphite cylinder (1 x 5 cm; Figure
7.20A) is designed to accommodate a much larger
sample aliquot—up to 100 µl—than is possible with
the filament and rod atomizers described previously.

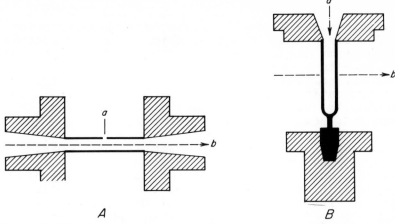

Figure 7.21. Arrangement of graphite tube and terminals for
the Massmann atomization furnace. (See
References 159, 197.)
A. For AA analysis
  ᵃSample port
  ᵇSource direction
B. For AF analysis
  ᵃSource direction
  ᵇFluorescence signal

In the earlier models, organic solutions tended to
migrate within the tube, but these have been rede-
signed with a grooved reservoir for use with such
solvents. Preatomization temperature programming,
much as previously described for the filaments, is
available also with the Perkin-Elmer furnace. But
Findlay *et al.*[199] have reported yet another example
of significant losses of, in particular, cadmium
and lead during the "ash" cycle with complex organic
matrix solutions, and it must be reemphasized that
all procedures involving preatomization heating of
the sample must be very carefully researched prior
to routine use. Cylinder wall temperatures up to
2700°C may be reached and, although the temperature
at the center of the tube will be several hundred
degrees cooler, this should be sufficient to atomize
most of the heavy metals with a high degree of
efficiency. Potentially troublesome radiation from
the incandescent graphite and electronic interferences
from the heating element may be minimized by suitable
optical baffling and magnetic shielding.[200] An inert
atmosphere must be maintained within the atomization
cell in the usual fashion, but it has been demonstrated,[201] not unexpectedly, that higher intensity

analysis signals result if the purge flow is momen-
tarily stopped during atomization, and this feature
is now provided automatically on the HGA-2000 model.

As noted previously, it would be no easy task
to quantitatively describe the AA signal in terms
of atomization rates and residence times within this
particular cell. Little specialized knowledge is
required, however, to appreciate which design and
operating parameters need be optimized for high
sensitivity analysis. For example, from Equation
3.24, it may be seen that the absorbance (A) is a
function both of the density of free atoms within
the cell (N) and the length of the cell tube (b),
so that, for this furnace, A_{max} improves as the
carbon tube diameter is decreased and the length
increased.[202]   A longer tube will also increase the
residence time of the atomic populations within the
cell. It is clear, of course, that there are practical limits imposed on the length of the furnace,
but, in any case, the various interference problems
considered below are simultaneously enhanced.

Depressive interference effects upon the absorption signal may be due, as with the other flameless
atomization modules, to variable atomization effi-
ciencies between certain samples and standards, or
to differences in the thermal regime along the length
of the tube. Furnaces are, however, more prone to
enhancements *via* light-scattering by particulates
or molecular band absorptions. These latter effects
may be corrected to varying degrees by employing a
compensating deuterium arc continuum in double-beam
type instrumentation. Kahn[202] has described the
application of this technique to the Perkin-Elmer
spectrometers. Similar corrections can be applied
when using single-beam instruments by making separate
measurements with a hydrogen source-lamp, but this
procedure is more time consuming and the degree of
compensation is likely to be considerably poorer,
particularly for resonance lines greater than, for
example, 3000 Å, since the latter source is far less
intense than the arc.[204]   The interference effects
observed in the presence of many matrices are, of
course, the result of so many contributing physicochemical reactions that it is next to impossible to
adequately compensate for these by purely instrumental means (see, for example, the work of Cruz
and Van Loon[205]). This has been well illustrated
by various studies which have attempted to determine
particular trace metals in sequence by differential
(programmed) atomization from concentrated matrices
such as seawater. Segar,[206,207] for example, has

shown that cadmium (which should be atomized at
temperatures below that required for the major sea
salt species) is retarded, whereas more "refractory"
elements such as copper are released at lower
temperatures than expected, presumably by some
co-volatilization mechanism. At our present state
of knowledge, it is advisable, wherever possible,
to chemically separate the matrix before attempting
AA analysis of the constituent trace metals.

Comparison of Nonflame Atomizers for AA

From the foregoing discussions, it should be
clear that no one particular flameless atomizer is
superior in every respect for the AA analysis of
trace heavy metals in water. Each analyst's samples
will be unique in some respect and actual choice of
instrumentation will be governed by this, together
with a host of personal circumstances. The best
that can be attempted here is a brief summary of
some of the more salient features of the more widely
distributed units. It is important to appreciate
that, despite some manufacturers' pronouncements to
the contrary, the detection limits of all flameless
AA techniques are only marginally suitable for the
direct determination of trace heavy metals in *most*
types of natural waters. This is, however, somewhat
immaterial since it is strongly advisable, for the
most precise analyses, to chemically separate the
metals of interest from either complex or concentrated matrices prior to the spectrometric analysis,
and it is a simple matter to incorporate at least a
small concentration step at this time.

West-Type Filament

This is useful if only very small sample volumes
are available, and it is recommended for the analysis
of organic extracts from natural waters. Very careful attention must be paid to the relative positions
of the reservoir and source beam. A fast response-
time detection system is advantageous. The available
thermal energy is less than with furnace-type units.
(This atomizer is very suitable for AF analysis.)

Carbon Rod Atomizer

This module (*i.e.*, the furnace-type variety of
the Varian Techtron Model 63 unit) exhibits higher

absolute sensitivities than the Massmann-type furnace,[208] but relative solution detection limits are comparable since the reservoir capacity is less. Automatic background correction is not possible with standard Varian Techtron equipment, but the background absorption effects are less troublesome than with the conventional furnace designs.

Tantalum-Strip Atomizer

This is recommended only for those metals which atomize at relatively low temperatures--silver, cadmium, lead, zinc and a few others. A double-beam optical system is incorporated in the IL instrument package.

Woodriff Furnace

This furnace is of an awkward size for use with most spectrometers, but the high temperatures generated ensure very efficient atomization of the more "refractory" heavy metals. Since atomization may be effected under essentially isothermal conditions, this design is eminently suitable for theoretical studies concerning the shapes of absorption signals *versus* atomization rates and residence times of atoms within the cell. A continuous sampling mode of operation is possible.

Massmann-Perkin-Elmer Furnace

This unit (the heated graphite analyser) can accommodate relatively large sample aliquots, but organic solutions require a modified tube configuration. It yields better overall analytical precision than the furnace-type rod atomizer.[208] Severe additive absorption interferences are commonly encountered, so that use of the automatic, continuum-source compensatory equipment provided (or available) on Perkin-Elmer instrumentation is virtually mandatory.

PLASMAS

Over the previous decade there has been considerable interest expressed in the potentialities of using plasmas as excitation sources for AE or as

atomizers for AA analysis. Plasmas offer several
theoretical advantages for these operations as compared with flame burners in a number of cases.
First and foremost, the thermal energy available is
considerably in excess of that given by the hottest
chemical flames. Even though "temperature" has
little exact meaning here since thermal equilibrium
within the plasma is not to be expected, the energy
available for excitation may be in the region of
10,000 K, and that for atomization not considerably
less. This means that even the most stable monoxides
(or other refractory complexes) are efficiently
dissociated, and also that solid samples may be
introduced for direct AS analysis. Samples are,
additionally, atomized into an essentially inert
atmosphere, so that the formation of intermediate,
difficult-to-dissociate compounds is inhibited. It
has been previously shown that this latter feature
is common also to the various nonflame atomizers,
but plasmas are also capable of accepting nebulized
sample; this mode is, in fact, more frequently
employed than discrete sampling. Plasmas exhibit
very low extraneous background radiation and are
optically "thin" emitting sources so that the
analysis concentration range for any one element
may extend over several orders of magnitude.

On the debit side, the required equipment is
complex and expensive. While various types of
plasma generation equipment are available on the
market, most of the developmental work cited in
these pages, particularly as applied to AA analysis,
has been accomplished using custom assembled instrumentation. There are also innumerable difficulties
associated with injecting sample (especially solutions) into the reaction cell. Some detection
limit data for both AA and AE analysis have been
included in Tables 3.5, 3.6 and 7.7. Application
to AA is clearly limited to elements which do not
readily ionize.

Arc Plasmas

Plasmas generated within a dc arc were early
investigated as AE excitation sources, and this
type necessitates possibly the least complex equipment. Several units, such as that illustrated in
Figure 7.22,[210] are available commercially. Various
procedures have been adopted to lift the reaction
cell away from the arc. The bend in the plasma
shown in Figure 7.22 is produced without resorting

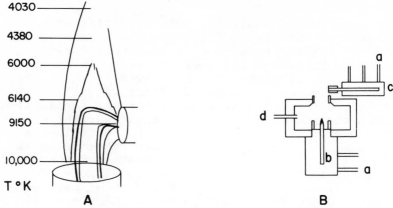

Figure 7.22.   A dc discharge plasma jet for high temperature
AE analysis. (After Reference 210 by courtesy
of International Scientific Communications Inc.)
A. Temperature profile
B. Basic construction
      *a*Argon inlet
      *b*Anode
      *c*Cathode
      *d*Sample inlet.

to external magnetic or electrical fields, and Avni
and Miron[211] have described a vacuum method for
extruding the plasma. Sample may be introduced by
direct pneumatic nebulization. Vickers,[212] for
example, has utilized two gas streams: one to
stabilize the arc, and the other to transport
sample aerosol.

Application of arc plasmas to AA analysis is
rendered difficult because of the short path length
of the absorption cell. This high temperature
method of atomization is obviously only really useful
for those metal compounds (*e.g.*, of tungsten, vanadium,
titanium, molybdenum, etc.) which are difficult to
dissociate in flame and nonflame devices.[213] Marinkovic
and Vickers[214] have described a long-path arc (around
9,000°K) better suited to AA analysis. These authors
have noted that monoxide formation was not completely
eliminated so that chemical additions designed to
encourage a reducing environment were advantageous.

Low-Power Microwave Plasmas

Low-wattage, microwave induced plasmas utilize
much the same generator and cavity equipment as

previously described for electrodeless discharge
tubes. These high electron-temperature devices[215]
are very suitable excitation sources for AE analysis
and require considerably less energy and insulation
of internal components than do the r.f. plasma
torches considered below. Unfortunately, a stable
plasma cannot be maintained if sample is introduced
in liquid form. In fact, because of the thermal
disequilibrium environment, the thermal energy
available for vaporization is quite low, so that
various ancillary means of accomplishing this step
have been devised. This requirement essentially
limits the application of low-wattage plasmas to
discrete samples vaporized from, for example, an
auxiliary arc[216] or the heated metallic loop.[162]
AE detection limits by this latter procedure are
cited in Table 7.7.

High-Power r.f. Plasmas

Early examples of the use of r.f. generated
plasmas as AE excitation sources were given by
Mavrodineanu and Hughes[217] and West and Hume,[218]
and Fassel and co-workers[219,220] have used inductively-
coupled torches for the determination of a range of
heavy metals by both AE and AA (see Tables 3.5 and
3.6). As with all plasmas, the principal practical
problem encountered has been that of sample introduction. In the case of the induction plasmas, for
example, skin heating effects tend to cause reflec-
tion of the sample particles. Ultra-sonic nebulization has invariably been employed. A comprehensive
study published in 1966[9] concluded that plasma
torches (Figure 7.4b) offered few advantages over
chemical flames for either AA or AE analysis. In
spite of the clearly superior thermal energy available, these authors cited in particular the numerous
operating inconveniences and the cost and complexity
of the equipment. Since this time, few workers have
advocated the use of plasma torches for AA analysis.
Very high temperatures *per se* tend to be disadvantageous for this mode of analysis because of
ionization problems, and other desirable features,
such as a favorable redox environment, have been
largely preempted by the various flameless atomization units. However, recent design improvements
have enhanced the value of these plasmas as emission
sources. Dickinson and Fassel[221] (see Table 3.6),
for example, have produced a torroidal shaped plasma
which facilitates sample addition, and AE detection

limits in the 0.001-0.02 µg/mℓ range for silver, arsenic, cadmium, copper and lead in a range of matrices have been given[222] using this equipment.

REFERENCES

1. Herrmann, R. in *Flame Emission and Atomic Absorption Spectrometry*, Vol. 2 (J. A. Dean and T. C. Rains, eds.), Marcel Dekker, New York, 1969, pp. 151-187.
2. L'vov, B. V. *Atomic Absorption Spectrochemical Analysis* (Translated from Russian), American Elsevier, New York, 1970.
3. Sacks, R. D. and V. T. Cordasco. 4th Intern. Conf. Atom. Spectrosc., Toronto, November 1973.
4. Gough, D. S., P. Hannaford and A. Walsh. *Spectrochim. Acta, 28B,* 197 (1973).
5. Walsh, A. *Appl. Spectrosc., 27,* 335 (1973).
6. Prudnikov, E. D. 4th Intern. Conf. Atom. Spectrosc., Toronton, November 1973.
7. Segar, D. A., J. G. Gonzalez, J. L. Gilio and R. E. Pellenbarg. 4th Intern. Conf. Atom. Spectrosc., Toronto, November 1973.
8. Lord, D., G. Van Loon and W. G. Breck. Proc. Symp. Water Anal. Parameters, Burlington, Ontario, November 1973.
9. Veillon, C. and M. Margoshes. *Spectrochim. Acta, 23B,* 553 (1968).
10. Venghiattis, A. A. *Spectrochim. Acta, 23B,* 67 (1967).
11. Harrison, W. W. and P. O. Juliano. *Anal. Chem., 43,* 248 (1971).
12. Alkemade, C. T. J. in *Flame Emission and Atomic Absorption Spectrometry*, Vol. 1 (J. A. Dean and T. C. Rains, eds.), Marcel Dekker, New York, 1971, pp. 101-150.
13. Gaydon, A. G. and H. G. Wolfhard. *Flames: Their Structure, Radiation and Temperature*, 3rd ed., Chapman and Hill, London, 1970.
14. Friston, R. M. and A. A. Westenburg. *Flame Structure*, McGraw-Hill, New York, 1965.
15. Dvorak, J., I. Rubeska and Z. Rezac. *Flame Photometry: Laboratory Practice* (English translation, R. E. Hester, ed.), CRC Press, Cleveland, 1971.
16. Mavrodineanu, R. *Develop. Appl. Spectr., 5,* 371 (1966).
17. Jenkins, D. R. and T. M. Sugden. in *Flame Emission and Atomic Absorption Spectrometry*, Vol. 1 (J. A. Dean and T. C. Rains, eds.), Marcel Dekker, New York, 1969, pp. 151-187.
18. d'Silva, A. P., R. N. Kniseley and V. A. Fassel. *Anal. Chem., 36,* 1287 (1964).
19. Bailey, B. W. and J. M. Rankin. *Anal. Chem., 43,* 219 (1971).

20. Boling, E. A. *Spectrochim. Acta,* *22,* 425 (1966).
21. Rubeska, J. and B. Moldan. *Atomic Absorption Spectrophotometry* (English translation, P. T. Woods, ed.), Iliffe, London, 1969.
22. Mavrodineanu, R. *Develop. Appl. Spectr.,* *8,* 18 (1970).
23. Bratzel, M. P. and J. D. Winefordner. *Anal. Lett.,* *1,* 43 (1967).
24. Koirtyohann, S. R. in *Flame Emission and Atomic Absorption Spectrometry,* Vol. 1 (J. A. Dean and T. C. Rains, eds.), Marcel Dekker, New York, 1969, pp. 295-315.
25. Bailey, B. W. and J. M. Rankin. *Anal. Chem.,* *43,* 216 (1971).
26. Willis, J. B. *Spectrochim. Acta,* *23A,* 811 (1967).
27. Riley, J. P. and D. Taylor. *Anal. Chim. Acta,* *42,* 548 (1968).
28. Ali, S. A. and D. C. Burrell. *Pakistan J. Sci. Ind. Res.,* *12,* 506 (1970).
29. Denton, M. B. and H. V. Malmstadt. *Anal. Chem.,* *44,* 241 (1972).
30. Kirsten, W. J. and G. O. B. Bertilsson. *Anal. Chem.,* *38,* 648 (1966).
31. Copeland, T. R., K. W. Olson and R. K. Skogerboe. *Anal. Chem.,* *44,* 1471 (1972).
32. Pungor, E. in *Atomic Absorption Spectroscopy* (R. M. Dagnall and G. F. Kirkbright, eds.), Butterworths, London, 1970, pp. 51-71
33. Clampitt, N. C. and G. M. Hieftje. *Anal. Chem.,* *44,* 1211 (1972).
34. Clampitt, N. C., G. J. Bastiaans and G. M. Hieftje. 4th Intern. Conf. Atom. Spectrosc., Toronto, November 1973.
35. Eshelman, H. C. and J. Armentor. *Develop. Appl. Spectr.,* *1,* 190 (1962).
36. Koirtyohann, S. R. and E. E. Pickett. *Anal. Chem.,* *40,* 2068 (1968).
37. Jenkins, D. R. *Spectrochim. Acta,* *25B,* 47 (1970).
38. Bouckaert, R., J. D'Olieslager and S. De Jaegere. 4th Intern. Conf. Atom. Spectrosc., Toronto, November 1973.
39. Halls, D. J. and E. Pungor. *Anal. Chim. Acta,* *44,* 40 (1969).
40. Behera, S. and C. L. Chakrabarti. 4th Intern. Conf. Atom. Spectrosc., Toronto, November 1973.
41. Coker, D. T., J. M. Ottaway and N. K. Pradhan. *Nature Phys. Sci.,* *233,* 69 (1971).
42. West, A. C., V. A. Fassel and R. N. Kniseley. *Anal. Chem.,* *45,* 1586 (1973).
43. Winefordner, J. D., V. Svoboda and L. J. Cline. *CRC Crit. Rev. Anal. Chem.,* *1,* 233 (1970).
44. Pickett, E. E. and S. R. Koirtyohann. *Anal. Chem.,* *41,* 28A (1969).
45. deGalan, L. and G. F. Samaey. *Spectrochim. Acta,* *25B,* 245 (1970).

46. Willis, J. B. *Spectrochim. Acta, 26B,* 177 (1971).
47. Fassel, V. A., J. O. Rasmuson, R. N. Kniseley, and T. G. Cowley. *Spectrochim. Acta, 25B,* 559 (1970).
48. Smyly, D. S., W. P. Townsend, P. J. Th. Zeegers and J. D. Winefordner. *Spectrochim. Acta, 26B,* 531 (1971).
49. Dean, J. A. and J. C. Simms. *Anal. Chem., 35,* 699 (1963).
50. Rann, C. S. and A. N. Hambly. *Anal. Chem., 37,* 879 (1965).
51. Hambly, A. N. and C. S. Rann. *Flame Emission and Atomic Absorption Spectrometry,* Vol. 1 (J. A. Dean and T. C. Rains, eds.), Marcel Dekker, New York, 1969, pp. 241-265.
52. Chakrabarti, C. L., M. Katyal and D. E. Willis. *Spectrochim. Acta, 25B,* 629 (1970).
53. Chakrabarti, C. L., R. Pal and M. Katyal. *Anal. Chem., 43,* 1704 (1971).
54. Pungor, E. and I. Cornides. in *Flame Emission and Atomic Absorption Spectrometry,* Vol. 1 (J. A. Dean and T. C. Rains, eds.), Marcel Dekker, New York, 1969, pp. 49-99.
55. Buell, B. E. in *Flame Emission and Atomic Absorption Spectrometry,* Vol. 1 (J. A. Dean and T. C. Rains, eds.), Marcel Dekker, New York, 1969, pp. 267-293.
56. Mavrodineanu, R. and H. Boiteux. *Flame Spectroscopy,* John Wiley and Sons, New York, 1965.
57. Zacha, K. E. and J. D. Winefordner. *Anal. Chem., 38,* 1537 (1966).
58. West, T. S. in *Atomic Absorption Spectroscopy* (R. M. Dagnall and G. F. Kirkbright, eds.), Butterworths, London, 1970, pp. 99-126.
59. Martin, T. L., F. M. Hamm and P. B. Zeeman. *Anal. Chim. Acta, 53,* 437 (1971).
60. Kniseley, R. N. in *Flame Emission and Atomic Absorption Spectrometry,* Vol. 1 (J. A. Dean and T. C. Rains, eds.), Marcel Dekker, New York, 1969, pp. 189-211.
61. Allan, J. E. *Spectrochim. Acta, 17,* 467 (1961).
62. Sachdev, S. L., J. W. Robinson and P. W. West. *Anal. Chim. Acta, 37,* 156 (1967).
63. Lemonds, A. J. and B. E. McClellan. *Anal. Chem., 45,* 1455 (1973).
64. Dean, J. A. *Flame Photometry,* McGraw-Hill, New York, 1960.
65. Buell, B. E. *Anal. Chem., 34,* 635 (1962).
66. Pungor, E., K. Toth and I. K. Thege. *Z. Anal. Chem., 200,* 21 (1964).
67. Fassel, V. A., R. H. Curry and R. N. Kniseley. *Spectrochim. Acta, 18,* 1127 (1962).
68. Skogerboe, R. K., A. T. Heybey and G. H. Morrison. *Anal. Chem., 38,* 1821 (1966).
69. Bode, H. and H. Fabian. *Z. Anal. Chem., 170,* 387 (1959).
70. Smith, R., C. M. Stafford and J. D. Winefordner. *Anal. Chem., 41,* 946 (1969).

71. Herrmann, R. and C. T. J. Alkemade. *Chemical Analysis by Flame Photometry*, 2nd ed., John Wiley and Sons, New York, 1963.
72. Mansell, R. E. *Atomic Absorption Newsletter, 6*, 6 (1967).
73. Willis, J. B. *Appl. Opt., 7*, 1295 (1968).
74. Willis, J. B. *Nature, 207*, 715 (1965).
75. Hell, A. and S. G. Ricchio. *Flame Notes, 4*, 37 (1969).
76. Willis, J. B., J. O. Rasmuson, R. N. Kniseley and V. A. Fassel. *Spectrochim. Acta, 23B*, 725 (1968).
77. Mossholder, N. V., V. A. Fassel and R. N. Kniseley. *Anal. Chem., 45*, 1614 (1973).
78. Browner, R. F., R. M. Dagnall and T. S. West. *Anal. Chim. Acta, 46*, 207 (1969).
79. Fassel, V. A., R. B. Myers, and R. N. Kniseley. *Spectrochim. Acta, 19*, 1194 (1963).
80. Kniseley, R. N., A. P. d'Silva, and V. A. Fassel. *Anal. Chem., 35*, 910 (1963).
81. Veillon, C., J. M. Mansfield, M. L. Parsons and J. D. Winefordner. *Anal. Chem., 38*, 204 (1966).
82. Ellis, D. W. and D. R. Demers. *Anal. Chem., 38*, 1943 (1966).
83. Ellis, D. W. and D. R. Demers. in *Trace Inorganics in Water*, (R. A. Baker, ed.), American Chemical Society, Washington, D.C., 1968, pp. 326-336.
84. Fuwa, K. and B. L. Vallee. *Anal. Chem., 35*, 942 (1963).
85. McGee, W. W. and J. D. Winefordner. *Anal. Chim. Acta, 37*, 429 (1967).
86. Kahn, H. L. and J. E. Schallis. *Atomic Absorption Newsletter, 7*, 5 (1968).
87. Amos, M. D. and J. B. Willis. *Spectrochim. Acta, 22*, 1325 (1966).
88. Mansell, R. E. *Spectrochim. Acta, 25B*, 219 (1970).
89. Kirkbright, G. F., M. K. Peters and T. S. West. *Talanta, 14*, 789 (1967).
90. Kirkbright, G. F., M. K. Peters, M. Sargent and T. S. West. *Talanta, 15*, 663 (1968).
91. Stephens, R. and T. S. West. *Spectrochim. Acta, 27B*, 515 (1972).
92. Sturgeon, R. E. and C. L. Chakrabarti. 4th Intern. Conf. Atom. Spectrosc., Toronto, November 1973.
93. Kornblum, G. R. and L. de Galan. *Spectrochim. Acta, 28B*, 139 (1973).
94. Parsons, M. L. and P. M. McElfresh. *Appl. Spectrosc., 26*, 472 (1972)
95. Pickett, E. E. and S. R. Koirtyohann. *Spectrochim. Acta, 23B*, 235 (1968).
96. Kirkbright, G. F., A. Semb and T. S. West. *Spect. Lett., 1*, 97 (1968).
97. Winefordner, J. D. in *Atomic Absorption Spectroscopy*, (R. M. Dagnall and G. F. Kirkbright, eds.), Butterworths, London, 1970, pp. 35-50.

98. Christian, G. D. and F. J. Feldman. *Appl. Spectrosc.*, *25*, 660 (1971).

99. Goguel, R. *Spectrochim. Acta*, *26B*, 313 (1971).

100. Thomas, P. E. and M. D. Amos. *Resonance Lines*, *1*, 1 (1969).

101. Kirkbright, G. F. and T. S. West. *Appl. Opt.*, *7*, 1305 (1968).

102. Kirkbright, G. F., M. Sargent and T. S. West. *Atomic Absorption Newsletter*, *8*, 34 (1969).

103. Kirkbright, G. F., A. Semb and T. S. West. *Talanta*, *14*, 1011 (1967).

104. Kirkbright, G. F., A. Semb and T. S. West. *Talanta*, *15*, 441 (1968).

105. Browner, R. F. and D. C. Manning. *Anal. Chem.*, *44*, 843 (1972).

106. Hobbs, R. S., G. F. Kirkbright and T. S. West. *Analyst*, *94*, 554 (1969).

107. Amos, M. D., P. A. Bennett and K. G. Brodie. *Resonance Lines*, *2*, 3 (1970).

108. Kirkbright, G. F., M. Sargent and T. S. West. *Talanta*, *16*, 245 (1969).

109. Kirkbright, G. F., M. Sargent and T. S. West. *Talanta*, *16*, 1467 (1969).

110. Dagnall, R. M., G. F. Kirkbright, T. S. West and R. Wood. *Analyst*, *95*, 425 (1970).

111. Manning, D. C. *Atomic Absorption Newsletter*, *4*, 267 (1965).

112. Fleming, H. D. *Spectrochim. Acta*, *23B*, 207 (1967).

113. Cresser, M. S., P. B. Joshipura and P. N. Keliher. *Spect. Lett.*, *3*, 267 (1970).

114. Chester, J. E., R. M. Dagnall and M. R. G. Taylor. *Anal. Chim. Acta*, *55*, 47 (1971).

115. Rains, T. C. in *Flame Emission and Atomic Absorption Spectrometry*, Vol. 1 (J. A. Dean and T. C. Rains, eds.), Marcel Dekker, New York, 1969, pp. 349-379.

116. Bailey, B. W. and J. M. Rankin. *Spect. Lett.*, *2*, 159 (1969).

117. Rubeska, I. and B. Moldan. *Appl. Opt.*, *7*, 1341 (1968).

118. Moldan, B. in *Atomic Absorption Spectroscopy*, (R. M. Dagnall and G. F. Kirkbright, eds.), Butterworths, London, 1970, pp. 127-143.

119. Ando, A., K. Fuwa and B. L. Vallee. *Anal. Chem.*, *42*, 818 (1970).

120. Ando, A., M. Suzuki, K. Fuwa and B. L. Vallee. *Anal. Chem.*, *41*, 1974 (1969).

121. Iida, C. and M. Nagura. *Spect. Lett.*, *3*, 63 (1970).

122. Rubeska, I. 4th Intern. Conf. Atom. Spectrosc., Toronto, November 1973.

123. Hingle, D. N., G. F. Kirkbright and T. S. West. *Talanta*, *15*, 199 (1968).

124. Robinson, J. W. and P. J. Slevin. *American Laboratory*, *4*, 10 (1972).

125. Kirkbright, G. F. *Analyst, 96,* 609 (1971).
126. Campbell, W. C., J. M. Ottaway and B. Strong. 4th Intern. Conf. Atom. Spectrosc., Toronto, November 1973.
127. Amos, M. D. *American Laboratory, 4,* 57 (1972).
128. Veillon, C. 4th Intern. Conf. Atom. Spectrosc., Toronto, November 1973.
129. Fuller, C. W. *Anal. Chim. Acta, 62,* 442 (1972).
130. Maessen, F. J. M. J. and F. D. Posma. 4th Intern. Conf. Atom. Spectrosc., Toronto, November 1973.
131. Burrell, D. C., V. M. Williamson and M.-L. Lee. 4th Intern. Conf. Atom. Spectrosc., Toronto, November 1973.
132. Reeves, R. D., B. M. Patel, C. J. Molnar and J. D. Winefordner. *Anal. Chem., 45,* 246 (1973).
133. Kahn, H. L., G. E. Peterson and J. E. Schallis. *Atomic Absorption Newsletter, 7,* 35 (1968).
134. Emmermann, R. and W. Luecke. *Z. Anal. Chem., 248,* 325 (1969).
135. Luecke, W. and R. Emmerman. *Atomic Absorption Newsletter, 10,* 45 (1971).
136. Everson, R. T. and W. G. Schrenk. 4th Intern. Conf. Atom. Spectrosc., Toronto, November 1973.
137. Delves, H. T. *Analyst, 95,* 431 (1970).
138. Chao, T. T. and J. W. Ball. *Anal. Chim. Acta, 54,* 166 (1971).
139. Kerber, J. D. and F. J. Fernandez. *Atomic Absorption Newsletter, 10,* 79 (1971).
140. West, T. S. and X. K. Williams. *Anal. Chim. Acta, 45,* 27 (1969).
141. Anderson, R. G., I. S. Maines and T. S. West. *Anal. Chim. Acta, 51,* 355 (1970).
142. Amos, M. D. *American Laboratory,* August, 33 (1970).
143. Matousek, J. P. and B. J. Stevens. *Clin. Chem., 17,* 363 (1971).
144. Alder, J. F. and T. S. West. *Anal. Chim. Acta, 51,* 365 (1970).
145. Aggett, J. and T. S. West. *Anal. Chim. Acta, 57,* 15 (1971).
146. Burrell, D. C. *Trans. Am. Geophys. Union, 52,* 1253 (1972).
147. Johnson, D. J., T. S. West and R. M. Dagnall. *Anal. Chim. Acta, 66,* 171 (1973).
148. Parker, C. R., J. Rowe and D. P. Sandoz. *American Laboratory, 5,* 53 (1973).
149. Reeves, R. D., C. J. Molnar and J. D. Winefordner. *Analyt. Chem., 44,* 1913 (1972).
150. Burrell, D. C. and M-L. Lee. Proc. Symp. Water Qual. Parameters, Burlington, Ontario, November 1973.
151. Steele, T. W. and B. D. Guerin. *Analyst, 97,* 77 (1972).

152. Molnar, C. J., R. D. Reeves, J. D. Winefordner, M. T. Glenn, J. R. Ahlstrom and J. Savory. *Appl. Spectrosc.*, *26*, 606 (1972).
153. Cantle, J. E. and T. S. West. *Talanta*, *20*, 459 (1973).
154. Alger, D., R. G. Anderson, I. S. Maines and T. S. West. *Anal. Chim. Acta*, *57*, 271 (1971).
155. Anderson, R. G., H. N. Johnson and T. S. West. *Anal. Chim. Acta*, *57*, 281 (1971).
156. Glenn, M. T., J. Savory, S. A. Fein. R. D. Reeves, C. J. Molnar and J. D. Winefordner. *Anal. Chem.*, *45*, 203 (1973).
157. Amos, M. D., P. A. Bennett, K. G. Brodie, P. W. Y. Lung and J. P. Matousek. *Anal. Chem.*, *43*, 211 (1971).
158. Patel, B. M., R. D. Reeves, R. F. Browner, C. J. Molnar and J. D. Winefordner. *Appl. Spectrosc.*, *27*, 171 (1973).
159. Massmann, H. *Spectrochim. Acta*, *23B*, 215 (1968).
160. Kaiser, H. and H. Specker. *Z. Anal. Chem.*, *149*, 46 (1955).
161. Morrison, G. H. and Y. Talmi. *Anal. Chem.*, *42*, 809 (1970).
162. Aldous, K. M., R. M. Dagnall, B. L. Sharp and T. S. West. *Anal. Chim. Acta*, *54*, 233 (1971).
163. Bratzel, M. P., R. M. Dagnall and J. D. Winefordner. *Anal. Chim. Acta*, *48*, 197 (1969).
164. Gilbert, T. R. and D. N. Hume. *Anal. Chim. Acta*, *65*, 461 (1973).
165. Goulden, P. D. and P. Brooksbank. 4th Intern. Conf. Atom. Spectrosc., Toronto, November 1973.
166. Chu, R. C., G. P. Barron and P. A. W. Baumgarner. *Anal. Chem.*, *44*, 1476 (1972).
167. Brodie, K. G. and J. P. Matousek. *Anal. Chem.*, *43*, 1557 (1971).
168. Varian Techtron Pty Ltd., Techtron AA5 Operations Manual, 1971.
169. Hall, G., M. P. Bratzel and C. L. Chakrabarti. *Talanta*, *20*, 759 (1973).
170. L'vov, B. V. *Spectrochim. Acta*, *24B*, 53 (1969).
171. L'vov, B. V. in *Atomic Absorption Spectroscopy*, (R. M. Dagnall and G. F. Kirkbright, eds.), Butterworths, London, 1970, pp. 11-34.
172. Massmann, H. *Z. Anal. Chem.*, *225*, 203 (1967).
173. Woodriff, R., R. W. Stone and A. M. Held. *Appl. Spectrosc.*, *22*, 408 (1968).
174. Welz, B. and E. Wiedeking. *Z. Anal. Chem.*, *252*, 111 (1970).
175. Fernandez, F. J. and D. C. Manning. *Atomic Absorption Newsletter*, *10*, 65 (1971).
176. Perkin-Elmer Corporation. P. E. 403 Operating Manual, 1971.

177. Slavin, S., W. Barnett and H. L. Kahn. *Atomic Absorption Newsletter, 11,* 37 (1972).
178. Donega, H. M. and T. E. Burgess. *Anal. Chem., 42,* 1521 (1970).
179. Parker, C. R. *Water Analysis by Atomic Absorption Spectroscopy.* Varian Techtron, Springvale, Australia, 1972.
180. Chakrabarti, C. L., Y. E. Araktingi and G. Hall. 4th Intern. Conf. Atom. Spectrosc., Toronto, November 1973.
181. Araktingi, Y. E. and C. L. Chakrabarti. 4th Intern. Conf. Atom. Spectrosc., Toronto, November 1973.
182. R. W. Morrow and R. J. McElhaney. 4th Intern. Conf. Atom. Spectrosc., Toronto, November 1973.
183. Hwang, J. Y., P. A. Ullucci and S. B. Smith. *American Laboratory,* August (1971).
184. Hwang, J. Y., P. A. Ullucci, S. B. Smith and A. L. Malenfant. *Anal. Chem., 43,* 1319 (1971).
185. Hwang, J. Y., C. J. Mokeler and P. A. Ullucci. *Anal. Chem., 44,* 2018 (1972).
186. Ullucci, P. A., C. J. Mokeler and J. Y. Hwang. *American Laboratory, 4,* 63 (1972).
187. Maruta, T. and T. Takeuchi. *Anal. Chim. Acta, 66,* 5 (1973).
188. King, A. S. *Astrophys. J., 75,* 379 (1932).
189. L'vov, B. V. *Spectrochim. Acta, 17,* 761 (1961).
190. Woodriff, R. and G. Ranelow. *Spectrochim. Acta, 23B,* 605 (1968).
191. Woodriff, R. and D. Shrader. *Anal. Chem., 43,* 1918 (1971).
192. Woodriff, R. and R. W. Stone. *Appl. Opt., 7,* 1337 (1968).
193. Woodriff, R., B. R. Culver and K. W. Olsen. *Appl. Spectrosc., 24,* 530 (1970).
194. Robinson, J. W., D. K. Wolcott, P. J. Slevin and G. D. Hindman. *Anal. Chim. Acta, 66,* 13 (1973).
195. Loftin, H. P., C. M. Christian and J. W. Robinson. *Spect. Lett., 3,* 161 (1970).
196. Robinson, J. W., G. D. Hindman and P. J. Slevin. *Anal. Chim. Acta, 66,* 165 (1973).
197. Massmann, H. in *Flame Emission and Atomic Absorption Spectrometry,* Vol. 2 (J. A. Dean and T. C. Rains, eds.), Marcel Dekker, New York, 1971, pp. 95-110.
198. Manning, D. C. and F. Fernandez. *Atomic Absorption Newsletter, 9,* 65 (1970).
199. Findlay, W. J., A. Zdrojewski and N. Quicker. 4th Intern. Conf. Atom. Spectrosc., Toronto, November 1973.
200. Kahn, H. L., J. D. Kerber and D. C. Manning. *American Laboratory, 5,* 55 (1973).
201. Kahn, H. L. and S. Slavin. *Atomic Absorption Newsletter, 10,* 94 (1971).

202. Welz, B. and W. Witte. 4th Intern. Conf. Atom. Spectrosc., Toronto, November 1973.
203. Kahn, H. L. *Atomic Absorption Newsletter, 7,* 40 (1968).
204. Kahn, H. L. and D. C. Manning. *American Laboratory, 4,* 51 (1972).
205. Cruz, R. and J. C. Van Loom. 4th Intern. Conf. Atom. Spectrosc., Toronto, November 1973.
206. Segar, D. A. and J. G. Gonzales. *Atomic Absorption Newsletter, 10,* 94 (1971).
207. Segar, D. A. and J. G. Gonzales. *Anal. Chim. Acta, 58,* 7 (1972).
208. Koirtyohann, S. R. 4th Intern. Conf. Atom. Spectrosc., Toronto, November 1973.
209. Sirois, E. H. *Anal. Chem., 36,* 2389 (1964).
210. Elliott, W. G. *American Laboratory,* August (1971).
211. Avni, R. and E. Miron. 4th Intern. Conf. Atom. Spectrosc., Toronto, November 1973.
212. Vickers, T. J. 4th Intern. Conf. Atom. Spectrosc., Toronto, November 1973.
213. Friend, K. E. and A. J. Diefenderfer. *Anal. Chem., 38,* 1763 (1966).
214. Marinkovic, M. and T. J. Vickers. *Appl. Spectrosc., 25,* 319 (1971).
215. Runnels, J. H. and J. H. Gibson. *Anal. Chem., 39,* 1399 (1967).
216. Layman, L. R. and G. M. Hieftje. 4th Intern. Conf. Atom. Spectrosc., Toronto, November 1973.
217. Mavrodineanu, R. and R. C. Hughes. *Spectrochim. Acta, 19,* 1309 (1963).
218. West, C. D. and D. N. Hume. *Anal. Chem., 36,* 412 (1964).
219. Wendt, R. H. and V. A. Fassel. *Anal. Chem., 37,* 920 (1965).
220. Wendt, R. J. and V. A. Fassel. *Anal. Chem., 38,* 337 (1966).
221. Dickinson, G. W. and V. A. Fassel. *Anal. Chem., 41,* 1021 (1969).
222. Winge, R. K., V. A. Fassel, R. N. Kniseley and W. L. Sutherland. Proc. Symp. Water Quality Parameters, Burlington, Ontario, November 1973.

CHAPTER 8

INTERFERENCES

"... the word interference will bring to mind a
welter of contradictory statements and hasty
generalizations, based on inadequate measurements
and misunderstandings.... It is now safe to say that
essentially the same interferences occur in both
[AE and AA analysis], for the same reasons, but to
somewhat different extents."

                              E. E. Pickett and S. R. Koirtyohann[1]

The term "interference effect" includes every
phenomenon causing deviations from the theoretical
AS growth curves, and because of this, it is an exceedingly complex topic. Many of the practical
laboratory manuals available to the analyst deal
with these effects in an entirely too superficial
fashion, frequently in terms of blanket remedies
for eliminating, suppressing or compensating for
the more common and serious problems. However, time
spent in exploring this subject in some depth will
be amply repaid. This is particularly so for the
attainment of maximum accuracy and precision--an
elusive goal in trace analysis at best--since several
problems from very different sources may require
attention. Several excellent review texts (cited
below) are now available to aid the laboratory
chemist, but a certain lack of uniformity of nomen-
clature still unfortunately persists due mainly to
the inseparable interrelationships between the
physical and chemical processes. The neophyte is
urged to refer to Rubeska and Modan[2] for their
masterful compaction of the principles central to
absorption interference phenomena. Because of the
enhanced importance of spectral interference to AE,

most standard texts (*e.g.*, Reference 3) devote
considerable space to this topic.

The generally assumed inferiority of AE as compared
with AA and AF for trace heavy-metal analysis is
based almost entirely on problems with interference
effects, predominantly spectral interferences, since
the elements involved require relatively high
energies both to dissociate the compounds and to
excite the atoms. Under such circumstances, spectra
of considerable complexity result from the analyte,
concomitants and flame constituents. Flame emission
spectral interferences are well known and daunting
to most analysts. However, they have been long
studied, are well understood and can frequently be
accommodated in all but the most complex matrices.
Pickett and Koirtyohann[1] have suggested that in many
cases skillfully tackling these problems in the AE
analysis mode is better laboratory practice than
employing AA/AE and assuming the absence--or being
unaware--of interference from various other sources.
Without doubt, the most troublesome effects in practical analysis are of nonspectral, physicochemical
origin. These effects may be large, are frequently
erratic and are essentially identical in *all* flame
AS methods.

SPECTRAL INTERFERENCES

This type of interference is concerned with
radiation emission, absorption and detection without
specific reference to the analyte. It is, as noted,
a topic of paramount importance in AE analysis, since
the analysis line must be differentiated from the
adjacent line, band and ionic spectra of the concomitants, and also from flame stray light continua.
These phenomena, then, are basically a function of
the instrument band-pass width; the severity of the
problem varies with the nature of the analyte and
matrix complexity.

Atomic Emission

Lines

Atomic line interference is particularly severe
in AE analysis and may well preclude the determination
of heavy metals by this technique. Table 8.1 shows

Table 8.1

Atomic Emission: Major Spectral Interferents
(Minor Analysis Lines in Parentheses)

Element	Line (Å)	Interferent	Line (Å)	Reference
Ag	3280.7	Cu	3274.0	4
	(3382.9)	Ni	3381[a]	5
As	(2349.8)	Be	2348.6	5
Au	2676.0	OH	2609[b]	5
Bi	3067.7	OH	b	5
	(2898.0)	OH	b	5
Cd	(3261.1)	Sn	3262.3	5
		OH	3261[b]	5
	2288.0	As	2288.1	5
Co	3453.5	Ni	3452.9	5
			3458.5	5
	(3502)[a]	Ni	c	4
		Mn	3532.0	4
	(3873.1)	CH	3872[b]	5
	(3874.0)	MgOH	3877[b]	5
		Fe	3872.5	5
Cr	3578.7	Co	c	4
	4254.3	Co	4252.3	5
		CH	4249[b]	5
Cu	3247.5	Ni	3233.0	4
			3243.1	4
	3274.0	Ag	3280.7	5
		OH	3274.2	5
Fe	3719.9	MgOH	3719[b]	5
Ga	(4033.0)	K	4044.1	4
		Mn	4033.1	4
	4172.1	BO$_2$	4180[b]	5
In	4101.8	SnO	b	4
	4511.3	Mo	d	5
		BO$_2$	4530[b]	5
Mn	4030.8	Ga	4033.0	5
		CH	4034[b]	5
	(4033.1)	K	4044.1	4
		Ga	4033.0	4
Mo	3798.3	MgOH	b	5
		Ru	3798.9	5
Ni	(3414.8)	Co	c	4
	3524.5	Co	c	4
		Mn	c	4
	(3019.4)	Fe	c	4
		Cr	c	4
Pb	4057.8	Mn	4057[a]	5
		Cu, CuH	4062[b]	5
		CH	4060[b]	5

Table 8.1, continued

Element	Line (Å)	Interferent	Line (Å)	Reference
Pd	3404.6	Co	3405.1	5
Rh	3692.4	Sn	3091.4	5
Ru	(3726.9)	MgOH	3729[b]	5
	3728.0	MgOH	3729[b]	5
Sb	2598.1	Po	2596[b]	5
Sn	(3175.0)	OH	3175[b]	5
Te	2386[a]	Fe	c	6
Tl	3775.7	Ni	3775.6	5
		MgOH	3767	5
		BaO	b	4
		Fe	c	4
		Ni	3775.6	4
V	3185.4[e]	OH	3185[b]	5
Zn	(4810.5)	Sr	4811.9	5

[a]Doublet   [d]Continuum
[b]Band   [e]Triplet
[c]Series

some potentially troublesome interferents. These are lines of major intensity; innumerable weak lines may be important in certain cases (see Reference 7). It may be found advantageous to design a pre-analysis treatment scheme (Chapter 5) specifically to isolate the test element or to remove one major interfering constituent of the sample if AE analysis is considered to offer advantages in other respects. Ion lines may interfere at elevated temperatures (see the ionization energies listed in Table 3.3) but these effects may be suppressed by the addition of ionization buffers in the usual fashion.

Bands

Broad spectral band interferences may derive from sample matrix species or from flame chemicals as noted below. The more common molecular band heads are given in Table 8.2, with examples of heavy-metal analysis lines most likely to be affected. Clearly, the AE determination of heavy metals in most natural waters is likely to be impossible without prior treatment. Band interferences are likely to be encountered also from oxides produced within the flame (where $E_{D.Ox}$ exceeds 5; see Table 3.3) and from organic solvent species.

Table 8.2

Some Common Band Interferents
(Data from Reference 5)

Complex	λ Range	Examples of Test Metals Affected
MgOH	3600–4000	Fe 3720
CaOH	5430–6220	(Ba 5536)
(CuH)	4000–4700	Mn 4031
		Pb 4058
(BO$_2$)	4000–6700	MnO 5390
(SrOH)	6000–7000	(Li 6708)

Flame Background Emission

This has been considered in the previous chapter (see Figures 7.5 and 7.6). Correction procedures for this type of interference require a sequence of background scanning on either side of the analysis line (*e.g.*, Reference 8).

Continua

Flame continua are weak but problems may arise from coexisting high concentrations of, for example, an alkali metal.

Corrections

AE spectral interferences may be eliminated or suppressed to a large extent by a combination of:

1. pre-analysis separation of interferents or analyte
2. use of various buffers (*e.g.* Reference 7)
3. background scanning

4. judicious selection of wavelength, flames
   and solvents
5. optimization of monochromator resolution/
   band-pass system
6. pulsing sample into flame and utilizing a
   tuned detection system[9]
7. passing modulated radiation from the emission
   cell through a second flame into which has
   been nebulized a solution containing a high
   concentration of the interferent. The
   emission intensity of the latter is thus
   attenuated by absorption.[10,11]

Atomic Absorption

The few documented[12-16] (Table 8.3) spectral
inferences likely to be encountered in AA analysis
are not likely to cause undue difficulties. They
are cited frequently in the literature in order to
demonstrate that this analysis mode is not entirely
free of these effects as was at one time supposed.

Table 8.3

Atomic Absorption: Spectral Interferences
(Minor Analytical Lines in Parentheses)

Element	Line (Å)	Interferent	Line (Å)	Reference
Cu	3247.54	Eu	3247.53	13
Fe	(2719.03)	Pt	2719.04	13
Ga	(4032.98)	Mn	4033.07	14
Hg	2536.52	Co	2536.49	15
Zn	2138.56	Na	-	16
		Fe	2138.59	17

More serious are various nonspecific (molecular)
background absorption effects on the source lines.
Koirtyohann and Pickett,[18] for example, have cited
the case of nonatomic absorption of cadmium in a
sodium matrix. At the present time, increasing use
is being made of compensating continuum sources
(Chapter 6) used alternately with the line source
to detect and correct for such background effects.[19]

Flame emission interferences are excluded in AA analysis by means of the modulated source and tuned detector system universally provided in commercial spectrometers. West[20] has suggested the possibility of fluorescence line interference on absorption measurements, but any such effects would be very small with standard sources.

Atomic Fluorescence

Atomic line interferences in line source AF should be comparable with the AA effects noted above. However, there is a certain amount of interest in the development and use of continuum sources for AF. In these cases spectral interferences might be expected to pose more serious problems. These effects have been investigated by, among others, Cresser and West[4] and Ellis and Demers,[21] who have shown that while atomic interferences may be important (Table 8.4), the molecular emission interferences which are so troublesome in AE work have little effect in AF. Flame emissions may be eliminated through source modulation as with AA, but, in fact, with the low background hydrogen flames frequently employed this precaution may be found unnecessary.

Table 8.4

Atomic Fluorescence: Spectral Interferences with a
Continuum Source (Minor Analytical Lines in Parentheses)
(Data from Reference 4)

Element	Line (Å)	Interferent	Line (Å)
Ag	3280.7	Cu	3274.0
Cd	2288.0	Ni	2320.0
Cu	3247.5	Ni	3233.0
In	4101.8	Co	4090.7
Mn	4033.1	K	4044.1
		Ga	4033.0
Ni	(3524.5)	Co	3526.8
Tl	3775.7	Fe	3737.1
			3734.9

PHYSICOCHEMICAL INTERFERENCES

This second type of interference is difficult to classify in any rational fashion mainly because of the near impossibility of separating the interacting effects of the contributing processes, many of which are, in any case, imperfectly understood. Included in this section are the conventional emission divisions of the physical (nonspecific) and chemical (specific) interferences (*e.g.*, Reference 22). However, it is felt that a chronological sequence[2]-- including effects attributable to transport of the sample *via* nebulization to the flame cell, and physicochemical processes primarily related to the condensed and vapor phase states of the sample, respectively--is preferable in several respects. It is appreciated that this system is open to considerable criticism, but it is deemed to satisfactorily serve the treatment given here. Each analyte, matrix and atomization procedure is subject to its own characteristic suite of interference problems and, unfortunately, only a very generalized treatment may be included here. The reader who feels a need for a more comprehensive and theoretical background in this subject is urged to consult the recent texts and especially References 23-25.

Transport Interferences

As defined in this book, sample transportation refers only to the movement of the sample to the flame cell or premix burner, so that these effects are absent when total consumption or nonflame cells are used. It must be remembered, however, that premix burners are designed in part to facilitate the subsequent sample atomization processes so that absence of transport effects does not necessarily impute any innate superiority to the latter devices. The transfer rate is a function of the particular physical parameters noted in Chapter 7 and, since there is no "ideal" transfer process, "interference" in this case refers to departure from the expected or standardized behavior. The analysis signal will vary with changes which affect the rate of flow, droplet/particle size or chemistry of the sample. Assuming that the internal components of the mixing chamber are left untouched, this type of interference should only be encountered with variations in the viscosity, temperature or density of the sample, the

aspiration rate, or the length or diameter of the
sampling capillary. One example of working curve
deviations related to the surface tension, viscosity
and density of the sample has been documented by
Wanninen and Lindholm;[26] of these parameters,
viscosity was found to be of enhanced importance.
Use of a syringe pump is recommended for ironing out
minor flow rate fluctuations, as noted previously.

Condensed and Vapor Phase
Interferences

This category of interference--primarily attributable to the nonideality of the conversion of sample
species to free atomic vapor--refers to reactions
occurring within the atomization cell. Irregularities
in the sample vaporization processes are considerably
more pronounced in flames, since the energy provided
is often insufficient or barely sufficient and the
reactions must be completed within the usable portion
of this dynamic cell. High temperature (enclosed)
nonflame cells are superior in both these respects
(but see Reference 27).
    Alkemade[28] has noted that condensed phase inter-
ferences may be considered in terms of:

1. signal enhancements or depressions due to the
   formation of compounds more or less volatile
   than the analyte
2. depressive or enhancement effects attributable
   to occlusion or dispersion of the analyte in a
   matrix material which is less or more easily
   volatilized.

It is, however, difficult in practice to separate
the resultant effects, and both types of effects may
occur for any given analyte and interferent. Condensed phase interferences of type (1) have long
been considered responsible for such classic examples
as the interaction of metal and phosphate radicals
to give nonvolatile phosphate compounds, and in the
ideal case, a relationship between the degree of
interference and some simple ratio of the reacting
chemical species might be expected. The virtual
impossibility of ascribing real signal discrepancies
to one particular cause has been well stated by
Fassel and Becker,[29] and such factors as the crystal
structure of the refractory compound may be important.[30] In the general case it should be noted that
interferences dependent upon the formation of some

specific refractory compound may occur where only
minor quantities of the interfering species are
present. Popham and Schrenk[31] have shown that the
depressive effects of several distinct species of
calcium (*e.g.*, Ca*, Ca°) are similar, suggesting
that dissimilar equilibria shifts are not the prime
cause. Neither did the mutual aluminum-calcium
interference, as a further example, appear to be
due to formation of a simple stoichiometric compound.
These workers concluded that occlusion of calcium
species in an aluminum compound matrix occurred
with sample evaporation, that is, a type (2)
interference as given above.

   Vapor phase interference reactions are equally
difficult to characterize, especially in flames
where the physical environment changes rapidly and
the presence of the flame reactant species adds to
the matrix complexity. A considerable portion of
the preceding chapter has been devoted to methods
of combating these effects, notably with regard to
the formation of oxide compounds. Other common
remedial techniques for a range of physicochemical
interferences depend upon the use of various chemical
additives to the sample[3,8,22,25] Fluctuations in
the temperature of the atomizing environment may
seriously affect emission intensities. One such
example concerning temperature quenching accompanying
the addition of aqueous sample to plasma torches has
been cited previously.

   Radiationless quenching of fluorescence emission
(Equation 3.16) is an exceedingly important type of
interference peculiar to this AS analysis mode.
Fluorescence yield values (y) of the type given in
Table 3.1 are, unfortunately, not available for many
metals because of the paucity of the necessary
quenching cross section information.[32,33] Sample
matrix concomitants have not been shown[34] to
specifically decrease the fluorescence intensity,
and problems in this area are virtually confined to
the flame combustion constituents. Of these,
monatomic species create the least problems, so that
nonhydrocarbon flames, and flames in which nitrogen
is replaced by, for example, argon (see Chapter 7)
are to be preferred for AF analysis. The effect of
OH radicals is not well known but Alkemade[35] has
noted that water may interfere *via* the dissociative
reaction:

$$R°* + H_2O \rightarrow R° + H + OH \qquad (8.1)$$

Clearly nonflame cells provide a less efficient quenching environment than do flames. Winefordner[36] has suggested that the resonance intensity from a graphite cell should exceed that from common flames by around an order of magnitude, other factors being equal.

REFERENCES

1. Pickett, E. E. and S. R. Koirtyohann. *Anal. Chem.*, *41*, 28A (1969).
2. Rubeska, J. and B. Moldan. *Atomic Absorption Spectrophotometry* (English translation, P. T. Woods, ed.), Iliffe, London, 1969.
3. Herrmann, R. and C. T. J. Alkemade. *Chemical Analysis by Flame Photometry*, 2nd ed., John Wiley and Sons, New York, 1963.
4. Cresser, M. S. and T. S. West. *Spectrochim. Acta*, *25B*, 61 (1970).
5. Buell, B. E. in *Flame Emission and Atomic Absorption Spectrophotometry*, Vol. 1 (J. A. Dean and T. C. Rains, eds.), Marcel Dekker, New York, 1969, pp. 267-293.
6. Dean, J. A. and J. C. Simms. *Anal. Chem.*, *35*, 699 (1963).
7. Dean, J. A. *Flame Photometry*, McGraw-Hill, New York, 1960.
8. Mavrodineanu, R. and H. Boiteux. *Flame Spectroscopy*, John Wiley and Sons, New York, 1965.
9. Rudiger, K., B. Gutsche, H. Kirchoff, and R. Herrmann. *Analyst*, *14*, 204 (1969).
10. Kirkbright, G. F. and S. J. Olson. *Spect. Lett.*, *2*, 225 (1969).
11. Kirkbright, G. F. and S. J. Wilson. *Analyst*, *95*, 833 (1970).
12. Koirtyohann, S. R. and E. E. Pickett. *Anal. Chem.*, *38*, 585 (1966).
13. Fassel, V. A., J. O. Rasmuson, and T. G. Crowley. *Spectrochim. Acta*, *23B*, 579 (1968).
14. Manning, D. C. and F. Fernandez. *Atomic Absorption Newsletter*, *7*, 24 (1968).
15. Allan, J. E. *Spectrochim. Acta*, *24B*, 13 (1969).
16. Kahn, H. L. *Atomic Absorption Newsletter*, *7*, 40 (1968).
17. Kelley, W. R. and C. B. Moore. *Anal. Chem.*, *45*, 1274 (1973).
18. Koirtyohann, S. R. and E. E. Pickett. *Anal. Chem.*, *37*, 601 (1965).
19. Sandoz, D. P. and D. L. Murray. *Resonance Lines*, *2*, 7 (1970).

20. West, T. S. in *U.S. National Bureau of Standards Monograph 100* (W. W. Meinke and B. F. Schribner, eds.), U.S. Gov't. Printing Office, Washington, D.C., 1967, pp. 121-148.

21. Ellis, D. W. and D. R. Demers. *Anal. Chem.*, *38*, 1943 (1966).

22. Gilbert, P. T. in Proc. 10th National Analysis Instrumentation Symposium, San Francisco, 1964, pp. 193-233.

23. Koirtyohann, S. R. in *Flame Emission and Atomic Absorption Spectrometry*, Vol. 1 (J. A. Dean and T. C. Rains, eds.), Marcel Dekker, New York, 1969, pp. 295-315.

24. Rubeska, I. in *Flame Emission and Atomic Absorption Spectrometry*, Vol. 1 (J. A. Dean and T. C. Rains, eds.), Marcel Dekker, New York, 1969, pp. 317-348.

25. Rains, T. C. in *Flame Emission and Atomic Absorption Spectrometry*, Vol. 1 (J. A. Dean and T. C. Rains, eds.), Marcel Dekker, New York, 1969, pp. 349-379.

26. Wanninen, E. and A. Lindholm. *Anal. Lett.*, *1*, 55 (1967).

27. Baudin, G., M. Chaput, and L. Feve. *Spectrochim. Acta*, *26B*, 425 (1971).

28. Alkemade, C. T. J. *Anal. Chem.*, *38*, 1252 (1966).

29. Fassel, V. A. and D. A. Becker. *Anal. Chem.*, *41*, 1522 (1969).

30. Sastir, V. S., C. L. Chakrabarti and D. E. Willis. *Talanta*, *16*, 1093 (1969).

31. Popham, R. E. and W. G. Schrenk. *Develop. Appl. Spectr.*, *7A*, 189 (1969).

32. Jenkins, D. R. *Spectrochim. Acta*, *23B*, 167 (1967).

33. Jenkins, D. R. *Spectrochim. Acta*, *25B*, 47 (1970).

34. West, T. S. in *Atomic Absorption Spectroscopy* (R. M. Dagnall and G. F. Kirkbright, eds.), Butterworths, London, 1970, pp. 99-126.

35. Alkemade, C. T. J. in *Atomic Absorption Spectroscopy* (R. M. Dagnall and G. F. Kirkbright, eds.), Butterworths, London, 1970, pp. 73-98.

36. Winefordner, J. D. in *Atomic Absorption Spectroscopy*, (R. M. Dagnall and G. F. Kirkbright, eds.), Butterworths, London, 1970, pp. 35-50.

CHAPTER 9

THE DATA

> "... there is an awful lot of monkey business
> being played with numbers, especially now that a
> penchant for being at least superficially quantita-
> tive has arrived on the environmental scene....
> Apart from the semantic mischief that can be wreaked
> on even accurate numbers ... the most dangerous
> aspect of numbers appears to be that they are rarely
> quoted with an indication of how they were measured."
> Editorial, *Env. Sci. Tech.*[1]

The analysis signals produced in the manner considered in Chapter 6 must be displayed in some
suitable fashion, recorded, and then evaluated. The
latter function is often neglected by the analyst--
indeed, very often, it is not formally his job--but
only he can fully appreciate the worth of the values
produced. Some of the operations and considerations
involved are briefly outlined in this chapter.

DATA PROCESSING

The methods used for displaying and interpreting
the analysis signals must be closely matched to the
type of AS analysis performed. This statement may
be thought self-evident, but it is surprising how
little consideration is given to these operations
in many cases. Of the relevant factors here,
the first and foremost is the *mode* of analytical
operation; in the way this term has been used
throughout this volume, this refers primarily to
either discrete or continuous sample introduction

Secondly the analyst must set some boundaries on
the required speed, precision and accuracy.
Lastly, routine samples of the same type are
amenable to very different processing as compared
with, for example, individual samples having widely
fluctuating test element concentrations and matrices.

Signal Readout

   Most commercial atomic spectrometers are equipped
with some form of dial readout which should be used
only as a rough guide to the pertinent signal range.
This form of display can only represent the signal
at a particular moment in time, and no assessment
of the systematic error variability may be made.
Some improvement in this respect is given by the use
of coupled digital readout devices. These usually
have the facility for averaging a preselected number
of individual readings which may improve the reliability of the final analysis value but does not set
the precision "boundaries." Such modules are primarily designed to facilitate routine analysis runs
with relatively unskilled operators, but may be
useful in cases where the analyte concentrations
may be closely circumscribed.
   By far the more preferable readout procedure for
most continuous-nebulization AS work is the analog
display afforded by a good strip-chart (*i.e.*, X-t)
recorder. These instruments give a permanent record
of the analysis signal (Figures 7.17 and 10.3) so
that the relevant statistical limits may be readily
determined. In addition, the experienced analyst
may learn a great deal of the details and problems
of specific analyses from the resultant signal
shapes (Ramirez-Muñoz[2] has given a useful, practical
atlas of AA and AE signal recorder shapes). For
flame analysis with continuous sample introduction
also, the separation of small signals from the background may often be aided through use of the scale
expansion facility commonly offered by the major
equipment manufacturers. Direct signal amplification
affects both the analysis signal and the background
noise, of course, so that some additional signal
smoothing ("noise suppression") may also be required.
These operations can improve reliability, but at the
expense of time and increased sample consumption,
so that the quantity of sample available may well
decide the final statistical limits sought.   The
nebulization period of automatic analysis using

rotating sample-changers should be matched to the
overall background noise expected for each batch of
samples.

The methods commonly employed for the detection,
measurement and display of discrete AS analysis
signals are currently the same as those for
continuous-mode operation, as discussed above. This
is hardly surprising in view of the fact that the
same instrumentation is normally used for both.
However, the signals produced by each of these
modes of analysis are quite distinct, and it is
unfortunate that the practical necessities arising
from these differences are not better appreciated.
Signals from discrete, nonflame atomization are
transient, not continuous, and clearly some integrated recording procedure is to be preferred over
one better suited to the "equilibrium" signals from
continuous nebulization. Only L'vov[3] appears to
have considered the practicalities of this topic in
any detail.

The integrated pulse area (Q_N) to be measured
may be defined:

$$Q_N = \int_0^{\tau_S} N_t(t) \, dt \tag{9.1}$$

Here N_t represents the total number of atoms within
the cell at the moment of time t, and τ_S is the
signal recording time which is assumed to be very
large. For the measurement of continuous signals,
where both t and the ratio τ_T/τ_C (τ_C is the residence
time of atoms *within* the cell and τ_T the duration
time for transfer *to* the cell) are very large, then
$dN_t/dt \rightarrow 0$ and

$$N_{equil} = \frac{\tau_c}{\tau_T} \cdot N_{tot} \tag{9.2}$$

But for the discrete, integrated signal:

$$Q_N = N_{tot} = \tau_c \tag{9.3}$$

The best detection method for such integrated signals
would be that of direct photon counting, as noted in
Chapter 6. Where X-t chart recorders are used it is

usually allowable (for most nonflame atomizers) to
record and measure peak heights (Figures 7.11 and
7.17), but this obviously depends upon τ_c. Many
cold-vapor mercury signals, for example, should be
integrated. Recorders for use with this type of
analysis must have very fast response times. It
has been shown[4] that the maximum signal-to-noise
ratio is obtained where τ_{cir} --the recording circuit
time constant--approaches τ_c.

The Working Curve

The final analysis values are obtained by relating
the intensity signals from the samples to those of
known (standard) concentrations *via* a working curve
concept. The theoretical aspects of AS growth
curves have been considered in Chapter 3 for various
limiting conditions, and it has been shown that, for
low concentrations, AE and AF emission intensities
are linearly related to the atomic concentrations
in the cell (but see below). The necessary background corrections to be applied to the AE signal
traces require considerable care,[5] however. In AA
analysis, the measured signal is a function of the
ratio I_A/I_0 (*i.e.*, of α) and, unlike AE and AF, is
usually the difference of two relatively intense
signals. From the derivation culminating in Equation
3.24, it may be seen that the atomic concentration
by this mode of analysis is a linear function of A
(the absorbance; note that A may be replaced by 2-log
% transmission), so that it is customary to prepare
AA working curves by initially converting instrumental
percentage absorption readings to A values. With
some commercial equipment this step may be conveniently
accomplished by using built-in logarithmic amplifier
circuits. However, as has been further noted in
Chapter 3, it is perfectly permissible, at low concentrations, to assume linearity between α and the
atomic concentration. There is, in fact, "no reason
to manipulate instrumental output when it is already
in the best form for analytical plots."[6]
Regardless of whether α or A is the parameter
plotted, it is often expedient to make use of rapid
electronic processing to compute the sample concentrations; but off-line operation, in preference to
the coupled mode noted above, is recommended. In
this fashion the analyst can intelligently control
the data input. As one example,[7] the required
portions of the analog trace may be machine digitized,
and final concentrations determined using relatively

simple programs. Figure 9.1 illustrates a computer plot of a carbon filament aqueous working curve for copper in the mg/ℓ range. Butler *et al.*[8] have exemplified the use of desk calculators to compute working curves, and many other examples[9-11] of the

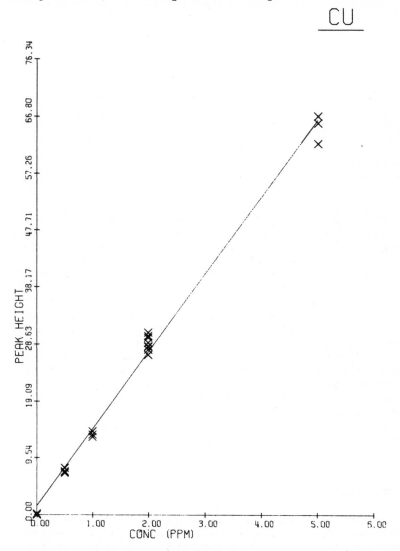

COPPER IN AQUEOUS SOLUTION ON WEST TYPE ROD
ONE MICROLITER SAMPLE SCALE EXPANSION = 1

Figure 9.1. Computer plot of aqueous working curve for copper via carbon filament atomization.

impact of such computer techniques are to be found
in the literature. Standard solutions must be treated
in the same fashion as the unknown samples, of course.
The correct use of standards to a large degree controls the final accuracy, as considered below.

Data Analysis

A comprehensive discussion of the best method
for statistical display of the data is beyond the
scope of this book, yet this is too important a
topic for the water chemist to pass completely un-
mentioned. The author is only too fully aware that
many analysts are not especially mathematically
inclined, but it behooves them to have a sufficient
understanding of the subject so that the essential
information may be fed to a nonchemically orientated
statistician. This point has been well emphasized
by Calder[12] in particular. A good introduction to
statistical analysis has been provided by Dixon and
Massey,[13] and Shaw and Bankier[14] have discussed the
application of the more useful techniques to geochemistry (some techniques of which are relevant
to water analysis). Mancy[15] has considered applicable
curve-fitting techniques.
In recent years, the methods of factor analysis
have found increasing use in the various geochemical
fields for determining correlations between multiple
parameters. Spencer[16] and Maxwell[17] have discussed
the principles involved. Application of these
specialized techniques presupposes collection of
useful ancillary data as noted in Chapter 4. For
aqueous metal pollutant data, this will require a
judicious selection of concomitant parameters.
Loder[18] has documented one application of the factor
analysis methods to natural water.

RELIABILITY

Data produced as discussed above is useless
without some estimation of the reliability which
may be placed upon the numbers. In order to accom-
plish this, it is imperative that the analyst be
aware of the total history of the sample up to and
including the final instrumental analysis. Inherent
errors may be considered in two categories--random
and systematic.

Nonsystematic errors--from a variety of causes--
are present in any batch of measurements, in such a
way that any one measurement (x) is randomly distributed about the true value x_O. If sufficient
measurements are taken, it will be seen that the
deviations $(x - x_O)$ are symmetrically dispersed
about the true value in the form of a "normal" or
Gaussian distribution. This is illustrated in
Figure 9.2 in which the frequency of observation or

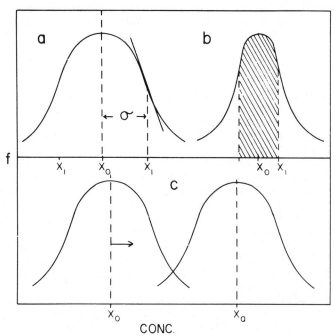

Figure 9.2. *Representation of analytical error assuming a*
Gaussian distribution. Frequency of occurrence
(f) *vs.* *concentration value; see text.*

a,b *Random error only, precision of (a) poorer*
than for (b).
c *Nonsystematic error.*

occurrence (f) is plotted against x:

$$f = \frac{\eta}{\sigma\sqrt{2\pi}} \cdot \exp\left[\frac{-(x - x_o)^2}{2\sigma^2}\right] \tag{9.4}$$

Here σ--the standard or rms deviation--is the distance between x_0 and x_i where x_i is the value of x at the inflection point. For n determinations:

$$\sigma = \left(\frac{\Sigma (x_i - \bar{x})^2}{n - 1} \right)^{\frac{1}{2}} \qquad (9.5)$$

where the mean ($\bar{x}$) is equal to x_0 if n is chosen large enough and systematic errors are absent. A more practical relative standard deviation (C) may be defined as a percentage parameter:

$$C = \frac{\sigma}{\bar{x}} \cdot 100 \qquad (9.6)$$

C is termed the coefficient of variation. The area beneath the frequency curve of Figure 9.2 within any given limits is a measure of the number of measurements or observations falling within these limits. Thus 68% of the measurements will fall within +σ of $\bar{x}$ (*i.e.*, of the true value x_0); this is the shaded area of Figure 9.2(b). Similarly 95.4% of the measurements--always assuming a sufficient number of observations are taken--lie within $\bar{x} + 2\sigma$, and 99.7% lie within $\bar{x} + 3\sigma$. Thus it is possible to assign confidence limits to the data. Reliability may be assigned at the 95 or 99% confidence level or, in the general case, measurement values may be assumed erratic if they fall outside $\bar{x} + k\sigma$ where the value of k depends upon the particular confidence level desired. Table 9.1 reproduces the coefficients of Dvorak *et al.*[19] for calculating data confidence levels; $\pm$ (σ · coefficient) gives the deviation at the required level. It must be remembered that this latter value is dependent upon the *number* of measurements taken (*i.e.*, the degrees of freedom) as emphasized above. The standard deviation is a function of the sharpness of the frequency curve; the smaller the value of σ, the better the reproducibility or precision. Thus in Figure 9.2, the precision of (a) is poorer than that of (b).

Systematic errors are nonrandom deviations or biases from the true or accurate datum x_0. In Figure 9.2(c) the difference ($x_0 - x_a$) is thus the magnitude of error where x_a is the measured mean value in the presence of one such bias. Systematic errors need not be present, but may be exceedingly difficult to detect, and constitute the major error problem in aqueous trace analysis for this reason. It should be noted that the precision describes

Table 9.1

*Coefficients of Dvorak, Rubeska and Rezac[19]
for Determining Precision Confidence Limits
(See text)*

| Number of | Confidence Level (%) | | |
Measurements	95.0	99.0	99.7
2	12.71	63.66	235.00
3	4.30	9.92	19.20
4	3.18	5.84	9.22
5	2.78	4.60	6.62
6	2.57	4.03	5.51
7	2.45	3.71	4.90
8	2.37	3.50	4.53
9	2.31	3.36	4.27
10	2.26	3.25	4.09
15	2.12	2.94	3.53
20	2.08	2.84	3.37
25	2.06	2.79	3.29
∞	1.96	2.58	2.97

the accuracy of the data only in such cases where nonrandom errors have been allowed for or are absent.

Pre-Analysis Errors

The importance of this topic cannot be over-emphasized. More error is likely to be introduced at this stage of the analysis scheme; nonrandom deviations will predominate. The latter may occur through loss or contamination addition during sampling or subsequent handling and processing of the sample.[20] Nonsystematic errors may be controlled by close attention to the design of the sampling program (as discussed elsewhere), but it is unlikely that systematic error may ever be entirely eliminated.

It is an essential practice to run processing and
reagent blanks through the entire sequence, and
radio-tracers may be employed to locate potential
areas of analyte loss (see Table 4.1), but this is
only fully worthwhile for routine and standard
techniques. At the very least, duplicate samples
should be taken from the water sampler and processed
and analyzed separately. But the sampler itself is
frequently a major source of distortion by addition
or removal. It is also clearly impossible to collect
an exactly duplicate water sample.

During the processing stage, the analyst must
be on guard both for ambient additive contaminants
and for irregular losses of the analyte. Paint
chips and foreign particulate matter (*e.g.*, from
the sampling vessel exhaust stacks) are prime culprits
at the sampling stage, and potential contamination
from filter membranes and various contacting surfaces
have been noted previously in Chapter 4 (see Tables
4.1-4.3). A negative bias *via* sorption of the test
metal onto the containing surfaces is a notoriously
common phenomenon with many metals (this is considered for silver as one example in the following
chapter). Additive contaminations from the reagent
chemicals used (Table 4.2) are equally frequently
encountered (see also Hume[21]).

Systematic errors may also be a consequence of
a poorly developed sampling program (Chapter 4),
which might inadvertently accommodate seasonal,
diurnal or water depth or locality biases. It is
imperative, as has been emphasized more than once
in these pages, that the data customer be thoroughly
familiar with the total history of the sample. In
many fields, the detection of systematic errors--
excluding those due to the initial sampling process--
may be accomplished by the use of standard materials
such as the U.S.N.B.S. clinical standards or the
standard rock powders issued by the U.S.G.S. No
such equivalent materials are available for the
water analyst, but a few interlaboratory and intertechnique comparative calibration studies have been
undertaken as referenced below.

Analytical Accuracy

Since AS is at present a comparative method of
analysis, analytical accuracy is entirely dependent
upon the correct selection, preparation and use of
the working curve standards.[22] Instructions for

the preparation of primary (1000 ppm) heavy-metal standard solutions are given in Appendix 3, but it is of paramount importance to matrix-match standards to the physical and chemical characteristics of the samples. Fewer problems of this type are likely to be encountered by the water analyst, however, than is the case in many other fields (in the analysis of trace constituents in dissolved silicate material, for example[23]). A coupled syringe pump may serve to iron out viscosity deviations as noted previously, and an ionization suppressant may be required for some elements for high temperature AA and AF analysis. Interference problems in general have been considered in Chapter 8. It should be standard practice to assess the need for continuum-source background correction (see Chapter 6) with AA procedures, and reagent blanks are a constant necessity. In case of special difficulties, or to check out a specific procedure, a method of standard additions should be employed. Routine application of the latter seriously slows analytical output of course, but the analyst must be resigned to the fact that the determination of trace and ultra-trace constituents in water is seldom a routine procedure. As one simple example, Figure 9.3 illustrates the use of a standard addition regression curve for the determination of copper in a seawater sample. This particular example, as

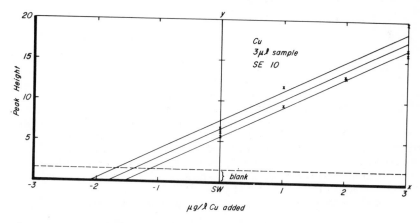

Figure 9.3. *Standard addition working curve for copper extracted into chloroform (x 10 concentration) from seawater using West-type carbon filament atomization.*

noted in Chapter 7, is for organic extracts (concentrated x 10) atomized on a West-type carbon
filament. Much of the computation drudgery associated with these operations may be alleviated through
use of flexible, multioptional electronic data
processing[7] and it is predominantly for this reason
that off-line over on-line processing was recommended above. A well-designed processing computer
program should also incorporate the facility for
recalculating the analytical curve frequently to
combat potential deviations in instrument response.
Special problems arise where organic solutions are
aspirated. It has, indeed, been only recently that
reliable organometallic standards have been commercially available. Some of these problems have been
considered by Hearn *et al.*[24]

The theoretical *shapes* of AS analytical (growth)
curves have been discussed in Chapter 3 (Figure 3.4),
where it was shown that the absorption, fluorescence
and emission intensities (I_A, I_F, I_E) are a linear
function of N_O at *low* atomic concentrations. It has
been further noted that the extended AF linearity
range[25,26] is a most useful analytical characteristic, and that--again at low concentrations--either
the absorbance (A) or the percentage absorption
(I/I_O) ratio may be plotted against concentration
to construct an AA working curve. In practice,
many factors operate to cause the working curve to
deviate from the theoretical growth curve (*i.e.*,
to bend). Variations in the degree of atomization
(β) due to variable ionization and dissociation
effects will promote marked curve bending, for
example, as will the detection of nonresonance
light--as a result of poor monochromator resolution
or some other optical defect--in AA analysis.[26,27]
This topic has been thoroughly aired in the
literature[27-31] and the interested reader is referred to the cited references for further information. It is sufficient to emphasize here that these
considerations will tend to dictate the working
curve range and the standard intervals employed by
the analyst. The use of the curve-straightening
facilities incorporated in some spectrometers is
not recommended for the type of high precision,
low level analysis to which this book is addressed.

Overall Precision

The standard deviation of the final analytical
data depends upon the errors accumulated at *every*
stage from sample retrieval through the various pre-
analysis treatments, in addition to the final instrumental analysis. Unfortunately, all too often
only the latter is considered when the reliability
of the data is assessed. Variances ($S = \sigma^2$) are
additive, so that the overall standard deviation
is given by:

$$\sigma = (\Sigma S)^{\frac{1}{2}} \qquad (9.7)$$

Random errors resulting from the pre-analysis
physical and chemical processing should be amenable
to reasonably close estimation by replicate analysis
in the usual fashion; the real villain here is nonrandom error as described above. For example,
Figure 9.4[32] shows recent data compiled from an

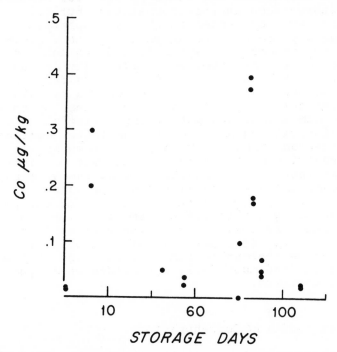

*Figure 9.4. Multiple laboratory determination of cobalt in
one seawater sample, concentration (µg/kg) vs.
time elapsed (days) between collection and
analysis. (Data from Reference 32. See text.)*

interlaboratory, trace-metal analysis (in seawater) calibration study. The mean of this particular set of data is 0.194 µg/Kg with a standard deviation of 0.129. However, these data are for samples which have been solvent extracted prior to analysis and the values are, in fact, considerably higher than those obtained for the same samples using direct analytical techniques. Assuming that the cobalt initially existed in the marine water in various complexed chemical forms, these results are the reverse of what might have been reasonably predicted. Obviously large systematic errors are involved. It might appear *a priori* that the pre-analysis treatment is the culprit, but this is not necessarily so in this example since the extracted samples were invariably analyzed by one method and the "totals" by another.

Considerable attention must be given to the statistical problems involved in primary sampling and subsampling programs. Various reasonably well-established techniques are commonly used for sampling static systems--such as rock formations--but the difficulties are compounded in the hydrosphere. Some standard procedures recommended by the U.S.G.S.[33] have been referenced in Chapter 4, and Venrick[34] has considered the errors involved specifically in sub-sampling for plankton population data. Calder[12] has demonstrated that less error is introduced by taking many samples and analyzing individually than by com-bining the subsamples and performing a single determination, although obviously the latter course is less time consuming. However, in attempting to improve precision by these means, one must be careful to guard against the introduction of nonrandom error. Water systems are not necessarily homogeneous and are dynamic to varying degrees, so that, for example, in a horizontally stratified system, important in-formation would be lost by combining samples from different layers. As repeatedly emphasized, the analyst must be thoroughly familiar with the system to be measured and must fully appreciate what type of data is required.

Analysis at Limiting Concentrations

The minimum usable analysis signal in AS is de-fined by the overall instrumental noise, so that the analysis detection limit is, in fact, a signal/noise ratio. Most quoted AS detection limit compilations

are with reference to twice the recorded background
noise (Figure 9.5).   The majority of the values
given in Tables 3.5 and 3.6 conform to this format,
but differing conventions are commonly applied to
other analytical techniques (see Table 2.13).
Limits defined thus are, however, somewhat unsatis-
factory since no indication of precision (according
to the principles discussed above) is included.

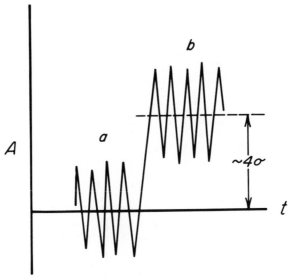

*Figure 9.5.   Idealized representation of continuous sampling
mode AS recorder trace (A or % absorption <u>vs</u>. t)
at the detection limit as commonly defined; see
text.*

*[a]Background (noise) signal.
[b]Analysis signal at 2x(a).*

     Comprehensive discussions on the theoretical
derivation of expressions for limiting detectable
concentrations in AS analysis have been given by
Winefordner and colleagues.[35-37]   The definition
employed by these workers is:

$$\frac{\text{Signal}}{\text{Noise}} = t \left(\frac{2}{n} \right)^{1/2}$$

(9.8)

where t is the Student "t" for a given confidence
level and n determinations.   This is a particular

form of the general expression for limiting detectability in chemical analysis.[38] The relationship advocated by Kaiser and Specker[39] is, however, in more widespread use. If $\Delta\bar{x}$ is the difference of $\bar{x}$ and $\bar{x}_b$--the mean of the blank (noise)--then:

$$\Delta\bar{x} = \text{Const} \cdot \sqrt{2\sigma_b} \qquad (9.9)$$

The constant of this equation depends upon the desired confidence level (*e.g.*, 3 for 99.7% certainty), so that $\Delta\bar{x}$ would approximate to $4\sigma_b$, which is also close to twice the background as shown in Figure 9.5. This relationship depends upon a reasonable approximation to a Gaussian distribution holding, that is, the inclusion of a sufficiently large number of observations (n of Equation 9.8). In this respect, the detection limit is a function of the quantity of sample analyzed or, for continuous nebulization, the duration of the sampling time. Figure 9.6(a) illustrates this correspondence for zinc. Detection

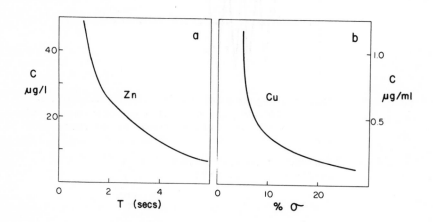

Figure 9.6. *Examples of limiting concentration (C) relationships.*

[a]Improvement of zinc detection limit with duration of aspiration (sampling) time; air-acetylene laminar-flow burner. (Data from Reference 40.)

[b]Dependence of analytical precision (%σ) on sample concentration range. (Data for copper using a laminar flow burner from Reference 41.)

limits of this type are, of course, relative to the
solution, whereas the discrete sampling mode compilations of Chapter 7 are absolute values. Analytical
precision decreases exponentially as the detection
limit is approached as shown in Figure 9.6(b). A
simple, practical procedure for determining AS
analysis limits has been published by Booher *et al.*[42]
    The term "sensitivity" was formerly, and con-
fusingly, used interchangeably with "detection limit,"
but it is preferable to reserve this term as a function
of the slope of the analysis curve. Ramirez-Muñoz[43],[44]
has employed the terms "fluctuational" and "percentual"
concentration limits in place of detection limit and
sensitivity respectively, but such terms as these
are an unnecessary proliferation and add little to
the analyst's tools. One definition for "sensitivity"
commonly used in the AA literature takes the units
of concentration/1% absorption as in Table 6.1.
Dean[45] has utilized the same concept--concentration/
1% transmission--for AE (see Table 6.3). These data
refer to continuous sampling mode analysis. Table
9.2 lists equivalent AA sensitivity values by nonflame
atomization; these data are *not* intercomparable.
    The above definition of sensitivity can only be
applied *sensu stricto* to a linear working curve,
although in practice this creates few problems since
the portion of the curve at limiting concentrations
is usually inferred. A more serious source of confusion may arise from the use of the reciprocal of
the slope of the working curve as conventionally
portrayed.[46] The sensitivity (γ) definition of
Mandel and Stiehler[47] corrects this discrepancy
[Figure 9.7(a)] and also takes into account the
analytical precision by incorporating a standard
deviation term:

$$\gamma = \frac{dI}{dC} \cdot \frac{1}{\sigma_I} \qquad (9.10)$$

where I represents the signal intensity and C is the
concentration, so that should two procedures produce
curves having the same slope; the one offering the
better precision is the more sensitive. Although
application of this expression to AS has been urged
by, especially, Skogerboe and Grant[48-50] it is in
little general use (but see Reference 51). Figure
9.7(b) illustrates the interdependence of the terms
of Equation 9.10; γ takes the units of reciprocal
concentration.

Table 9.2

Atomic Absorption Sensitivities (pg/1% absorption):
Discrete Sampling Mode

Element	Line (Å)	L'vov Crucible[a]	Welz/PE Furnace[b]	Woodriff Furnace[c]
Ag	3281	0.1		8
As	1937		200	
Au	2428	1	60	
Bi	3068[d]	4	300	
Cd	2288	0.08	1	9
Co	2407	2	100	
Cr	3579	2	20	
Cu	3248	0.6	50	90
Fe	2483	1		100
Ga	2874	1	1,000	
Hg	2537	80	20,000	
In	3039	0.4		
Mn	2795	0.2	7	80
Mo	3133	3		
Ni	2320	9	300	200
Pb	2833	2	20	10
Pd	2476	4	300	
Pt	2659	10	700	
Rh	3435	8		
Sb	2311[e]	5	500	
Se	1961	9		
Sn	2863	2	6,000	
Te	2143	1		
Ti	3653	40	300	
Tl	2768	1	90	
V	3184		300	
Zn	2139	0.03	2	9

[a]Extrapolations of absolute detection limits to 1% absorption; volume of sample used not defined (Reference 52).

[b]Data from References 53-55; 100-μl sample utilized.

[c]Data from Reference 56; 50-μl sample utilized.

[d]Line not utilized for flame absorption.

[e]Minor flame absorption line.

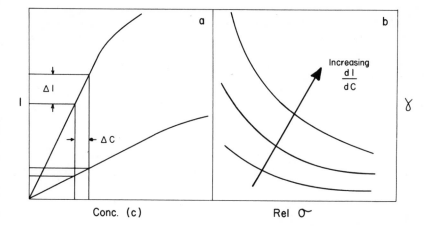

Figure 9.7. Analytical sensitivity as defined by Mandel and Stiehler.[47] *(See Equation 9.10; after Reference 57 by courtesy of Marcel Dekker Inc.)*

[a]*Sensitivity as slope (ΔI/ΔC) at fixed precision.*

[b]*Relationship between analytical precision as relative standard deviation (σ) of signal intensity (I), working curve slope, and sensitivity as defined.*

The above discussion (see also Reference 58) has been concerned with ideal concepts of analytical sensitivity and limiting concentrations. It is hoped that the detection limits for the heavy metals and various techniques quoted throughout this book will serve only as a guide to the potential suitability of a particular method and possibly as an indication of the reliability of the data thus obtained. West[26] has succinctly noted that "... the detection limit is a rather deceptive butterfly which should be chased warily. Chemistry, and in particular analytical chemistry, is a *practical* science based on quantification." It is, in short, very unwise to choose an analytical method solely or even largely on the basis of published detection limits, although this is the commonest of commercial blandishments. Many other factors, such as selectivity as a function of the overall sample chemistry, must be fully taken into account.

High Precision Flame Analysis

Assuming the absence of nonsystematic errors, increasing the reliability of an analysis involves optimizing the instrumental components[59] and operating parameters. In general, most of the instrumental variables (x) may be "peaked" with respect to the measured intensity, as illustrated schematically in Figure 9.8(a). This concept will

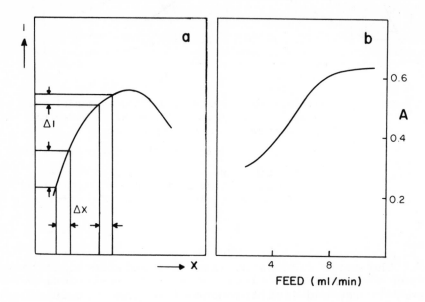

Figure 9.8.   *Optimization of instrumental parameters for high precision AF analysis.   (After Reference 57 by courtesy of Marcel Dekker Inc.)*

[a] *Generalized relationship between parameter (x) and measured intensity (I).*

[b] *Chromium absorbance (A) as a function of the rate of feed of sample (via a coupled syringe pump) to the nebulizer.*

be quite familiar to the average AS analyst with regard to such routine parameters as lamp currents, burner heights, etc., but a thorough grounding in the theoretical aspects of this topic (see References 60-64) will yield additional rewards. Figure 9.8(b) is one specific example showing the effect of sample feed rate on the absorbance of chromium.

Comparative Studies

There have been very few comparative studies of
different trace-metal methods or interlaboratory
calibration programs. Unfortunately, there are no
widely recognized heavy-metal water standards, so
nonrandom errors may easily pass undetected. Table
2.14[65] summarizes the results of one such experiment
which compared data by AA with that obtained by
several other analytical methods, but the coverage
is patchy and it is difficult to derive any definitive
feeling for the suitability of AA for a specific
sample type. Fishman and Mallory[66,67] have compared
four procedures--including AA--for molybdenum in
fresh water with comparable results and have also
looked at AA analysis *versus* the standard dithizone
method for zinc. The interlaboratory study of the
trace-metal contents of certain marine water samples
cited above[32] was, with regard to the AS techniques
utilized, more an evaluation of the pre-analysis
concentration chemistry.

UTILITY OF DATA

It is imperative to appreciate exactly what the
analysis numbers mean in terms of the original body
of water. The importance of knowing the total
processing history of the samples has been stressed
several times in the preceding pages and, since the
data will be eventually referred back to a more-or-
less dynamic water system, any limitations imposed
by the initial sampling program must be clearly
understood. It is important to know, for example,
to what extent the data may be extrapolated to
adjacent water layers or localities. There are
innumerable documented procedures for determining
the residence times and mixing rates of conservative
pollutants introduced into estuarine[68] or near-shore
freshwater areas.[69] However, such computations will
not--even with the incorporation of decay terms to
allow for the simpler nonconservative characteristics
of most chemical impingements--suitably model heavy-
metal distributions because of the multiplicity of
sorption and biotic uptake processes operating.
Chemical speciation of the trace-metal contents
has also been noted many times. Total soluble contents may be specifically sought, but this could
obscure important properties of the metals. For
example, reactive fractions may be included with
essentially immobilized forms. This problem is

analogous to the lithogenous and authogenous fractions of sediments. Frequently less-than-total soluble values are obtained haphazardly, and often inadvertently, as in cases where solvent extraction isolates an ill-defined "small complex" fraction. Table 2.2 is a good illustration of the need for special (although, in this case, relatively simple) treatment to reveal the presence of organically complexed metallic species.

The widespread sorption of heavy metals onto coexisting solid phases has been stressed. It has not been possible here to consider silicate analysis programs, but it must be noted that it is the sorbed (exchangeable) fraction rather than the structurally bound (lithogenous) content which is of enhanced importance. Yet it is also likely that many heavy metals are removed from solution to the particulates by processes which are largely irreversible. The same strictures apply to metals taken up by the aqueous biota. There is a world of difference between metals physiologically incorporated and those weakly bound to membrane surfaces, so that, again, the *total* trace metallic contents of these organisms may convey very little meaningful information. At the very least, the contents of the constituent parts of the animals should be differentiated as in Table 10.6, but it has been observed also that gradients may exist across single organs of the higher organisms.

The interpretation of the biotic heavy-metal contents in terms of toxicity/mortality is a field fraught with danger, and many of the "scare" statistics about which so much has been made in the popular literature are misleading at best. "Lethal concentrations" such as have been cited in Tables 2.5 and 10.6 should be treated with the greatest caution, as should familiar indices of the LD_{50} type. Toxicity varies from group to group and individual to individual and is a function of a whole range of ancillary water parameters such as hardness, temperature and dissolved oxygen content. The effect of one specific metal on an organism may even depend upon the character of the total coexisting trace-metal spectra.

Most fish heavy-metal poisons appear to act initially upon the gill epithelium, but there are a variety of pathways for the metals within the organism. The flushing rates and general dynamics of the water system are all important as noted by Lloyd.[70] A complementary field which has received minimal attention to date, but which is of the greatest importance,[71] is that of the role of heavy-metal toxicants present in solution in sublethal concentrations.

REFERENCES

1. *Environ. Sci. Technol.*, *6*, 1 (1972).
2. Ramírez-Muñoz, J. *Flame Notes*, *2*, 56 (1957).
3. L'vov, B. V. *Atomic Absorption Spectrochemical Analysis* (Translated from Russian), American Elsevier, New York, 1970.
4. L'vov, B. V. in *Atomic Absorption Spectroscopy* (R. M. Dagnall and G. F. Kirkbright, eds.), Butterworths, London, 1970, pp. 11-34.
5. Pickett, E. E. and S. R. Koirtyohann. *Anal. Chem.*, *41*, 28A (1969).
6. Zeegers, P. J. T., R. Smith, and J. D. Winefordner. *Anal. Chem.*, *40*, 26A (1968).
7. Burrell, D. C. *Anal. Chim. Acta*, *38*, 447 (1967).
8. Butler, L. R. P., P. F. S. Jackson, and K. Kroger. *Spect. Lett.*, *4*, 195 (1971).
9. Boling, E. A. *Anal. Chem.*, *37*, 482 (1965).
10. Scribner, B. F. *Appl. Spectrosc.*, *21*, 375 (1967)
11. Malakoff, J. L., J. Ramírez-Muñoz, and C. P. Aime. *Anal. Chim. Acta*, *43*, 37 (1968).
12. Calder, A. B. *Anal. Chem.*, *36*, 25A (1964).
13. Dixon, W. J. and F. J. Massey. *Introduction to Statistical Analysis*, McGraw-Hill, New York, 1957.
14. Shaw, D. M. and J. D. Bankier. *Geochim. Cosmochim. Acta*, *5*, 111 (1953).
15. Mancy, K. H. (Ed.) *Instrumental Analysis for Water Pollution Control*, Ann Arbor Science Publishers, Ann Arbor, Michigan, 1971.
16. Spencer, D. W. *Factor Analysis*, Clearinghouse for Federal Scientific and Technical Information Report AD 637 792, U.S. Dept. of Commerce, Washington, D.C., 1966.
17. Maxwell, A. E. *Factor Analysis as a Statistical Method*, Butterworths, London, 1971.
18. Loder, T. C. Ph.D. Dissertation, Institute of Marine Science, University of Alaska, College, Alaska, 1971.
19. Dvorak, J., I. Rubeska, and Z. Rezac. *Flame Photometry: Laboratory Practice* (English Translation, R. E. Hester, ed.), CRC Press, Cleveland, 1971.
20. Eisenhart, C. *Science, 160*, 1201 (1968).
21. Hume, D. N. in *Equilibrium Concept in Natural Water Systems* (R. F. Gould, ed.), American Chemical Society, Washington, D.C., 1967, pp. 30-44.
22. Schrenk, W. G. in *Flame Emission and Atomic Absorption Spectrometry*, Vol. 2 (J. A. Dean and T. C. Rains, eds.), Marcel Dekker, New York, 1971, pp. 303-325.

23. Burrell, D. C. *Norsk Geol. Tidsskr.*, *45*, 21 (1965).
24. Hearn, W. E., R. A. Mostyn, and B. Bedford. *Anal. Chem.*, *43*, 1821 (1971).
25. Syty, A. in *Flame Emission and Atomic Absorption Spectrometry*, Vol. 2 (J. A. Dean and T. C. Rains, eds.), Marcel Dekker, New York, 1971, pp. 197-233.
26. West, T. S. in *Atomic Absorption Spectroscopy* (R. M. Dagnall and G. F. Kirkbright, eds.), Butterworths, London, 1970, pp. 99-126.
27. van Gelder, Z. *Spectrochim. Acta, 25B*, 609 (1970).
28. Rubeska, I. and V. Svoboda. *Anal. Chim. Acta, 32*, 253 (1965).
29. Vickers, T. J., L. D. Remington, and J. D. Winefordner. *Anal. Chim. Acta, 36*, 42 (1966).
30. de Galan, L. and G. F. Samaey. *Spectrochim. Acta, 24B*, 679 (1969).
31. Roos, J. T. H. *Spectrochim. Acta, 25B*, 530 (1970).
32. Brewer, P. G. and D. W. Spencer. *Trace Element Intercalibration Study*, Woods Hole Oceanographic Institute, Technical Report No. 70-62, Woods Hole, Mass., 1970.
33. Rainwater, F. H. and L. L. Thatcher. Geol. Surv. Water Supply Paper 1454, U.S. Gov't. Printing Office, Washington, D.C., 1960.
34. Venrick, E. L. *Limnol. Oceanog., 16*, 811 (1971).
35. Winefordner, J. D. and T. J. Vickers. *Anal. Chem., 36*, 1939 (1964).
36. Winefordner, J. D. and T. J. Vickers. *Anal. Chem., 36*, 1947 (1964).
37. Winefordner, J. D., M. L. Parsons, J. M. Mansfield, and W. J. McCarthy. *Anal. Chem., 39*, 436 (1967).
38. Gabriels, R. *Anal. Chem., 42*, 1439 (1970).
39. Kaiser, H. and H. Specker. *Z. Anal. Chem., 149*, 46 (1955).
40. Ramírez-Muñoz, J. and M. E. Roth. *Flame Notes, 4*, 21 (1969).
41. Boettner, E. A. and F. I. Grunder. in *Trace Inorganics in Water* (R. A. Baker, ed.), American Chemical Society, Washington, D.C., 1968, pp. 236-246.
42. Booher, T. R., R. C. Elser, and J. D. Winefordner. *Anal. Chem., 42*, 1677 (1970).
43. Raimírez-Muñoz, J. *Talanta, 13*, 87 (1966).
44. Raimírez-Muñoz, J., N. Shifrin, and A. Hell. *Microchem. J., 11*, 204 (1966).
45. Dean, J. A. *Flame Photometry*, McGraw-Hill, New York, 1960.
46. de Galan, L. *Spect. Lett., 3*, 123 (1970).
47. Mandel, J. and R. D. Stiehler. *J. Res. Nat. Bur. Std., A, 53*, 155 (1954).
48. Skogerboe, R. K., A. T. Heybey, and G. H. Morrison. *Anal. Chem., 38*, 1821 (1966).
49. Grant, C. L. in *Atomic Absorption Spectroscopy*, Special Technical Publ. No. 443, American Society for Testing and Materials, Philadelphia, Pa., 1969, pp. 37-46.

50. Skogerboe, R. K. and C. L. Grant. *Spect. Lett.*, *3*, 215 (1970).
51. Massmann, H. *Spectrochim. Acta*, *23B*, 215 (1968).
52. L'vov, B. V. *Spectrochim. Acta*, *24B*, 53 (1969).
53. Fernandez, F. J. and D. C. Manning. *Atomic Absorption Newsletter*, *10*, 65 (1971).
54. Welz, B. and E. Wiedeking. *Z. Anal. Chem.*, *252*, 111 (1970).
55. Perkin-Elmer Corp., P. E. 403 Operating Manual, 1971.
56. Woodriff, R., R. W. Stone, and A. M. Held. *Appl. Spectrosc.*, *22*, 408 (1968).
57. Skogerboe, R. K. in *Flame Emission and Atomic Absorption Spectrometry*, Vol. 1 (J. A. Dean and T. C. Rains, eds.), Marcel Dekker, New York, 1969, pp. 381-411.
58. Menis, O. and T. C. Rains. in *Analytical Flame Spectroscopy: Selected Topics* (R. Mavrodineanu, ed.), MacMillan, London, 1970, pp. 47-77.
59. Rains, T. C. and O. Menis. *Spect. Lett.*, *2*, 1 (1969).
60. Crawford, C. M. *Anal. Chem.*, *31*, 343 (1959).
61. Ross, J. T. H. *Spectrochim. Acta*, *24B*, 255 (1969).
62. Parsons, M. L. and J. D. Winefordner. *Appl. Spectrosc.*, *21*, 368 (1967).
63. Parsons, M. L., W. J. McCarthy, and J. D. Winefordner. *J. Chem. Educ.*, *44*, 214 (1967).
64. McCarthy, W. J. in *Spectrochemical Methods of Analysis* (J. D. Winefordner, ed.), John Wiley and Sons, New York, 1971, pp. 493-518.
65. McFarren, E. F. and R. J. Lishka. in *Trace Inorganics in Water* (R. A. Baker, ed.), American Chemical Society, Washington, D.C., 1968, pp. 253-264.
66. Fishman, M. J. and E. C. Mallory, Jr. *J. Water Pollution Control*, *40*, R67 (1968).
67. Fishman, M. J. *Atomic Absorption Newsletter*, *5*, 102 (1966).
68. Bowden, K. F. in *Estuaries* (G. H. Lauff, ed.), Amer. Assoc. Adv. Sci., Washington, D.C., 1967, pp. 15-36.
69. Wnek, W. J. and E. G. Fochtman. *Environ. Sci. Technol.*, *6*, 331 (1972).
70. Lloyd, R. in Proc. 3rd Sem. on Biological Problems in Water Pollution, U.S. Dept. of Health, Education and Welfare, Washington, D.C., 1962.
71. Wilber, C. G. in Proc. 3rd Sem. on Biological Problems in Water Pollution, U.S. Dept. of Health, Education and Welfare, Washington, D.C., 1962.

CHAPTER 10

SOME MAJOR POLLUTANT METALS

"The quantitative determination of all minor
elements likely to be present in a [public] water
supply was, until recently, a forbidding undertaking.
The application of spectrochemical and atomic absorption techniques, however, has greatly facilitated
the task."

P. R. Barnett et al.[1]

In this chapter the current "state-of-the-art"
for the AS analysis of a few heavy metals of current
major concern as potential aqueous polluters will be
considered: *viz.* silver, arsenic, cadmium, mercury,
lead and selenium. These examples are intended to
be representative rather than exhaustive in any way,
and reference is made also to non-AS methods where
these offer comparable or superior advantages.

It is felt rather strongly that the analyst
should have, in addition to his specialized skills,
a reasonable feeling for the general environmental
chemistry of these metals. This is important for
several reasons. In the first place such knowledge
is a necessary prerequisite for the design of a
scientifically correct sampling program and for the
subsequent chemical treatment of these samples (as
considered in the preceding pages). But of equal
importance is the requirement for a balanced interpretation of the values obtained. It would seem
self-evident, for example, that the natural contents
and dynamics of heavy metals in biota should be well
understood *prior* to pronouncements concerning alleged
anthropogenic enhancements, but, unfortunately, this
has seldom been the case to date. Improper use of
the data in this fashion may result in unnecessary

economic stress, for example, in cases where mandatory
upper concentration limits for metals in food substances have been set on very arbitrary scientific
grounds. It is also true--contrary to widespread
opinion--that no well-documented examples of aqueous
food-web concentrations of these metals exist.

Table 10.1 shows recent statistics for the total
U.S. consumption of the "major" metal pollutants,

Table 10.1

Approximate Total U.S. Consumption of
Selected Pollutant Metals (Million Tons)
(Data from Reference 2)

Element	1948	1968	% Increase
Ag	0.004	0.005	40
As	0.024	0.025	4
Cd	0.004	0.007	70
Hg	0.002	0.003	60
Pb	1.134	1.329	17
Se	4×10^{-4}	8×10^{-4}	100
Zn	1.200	1.728	44

and Table 10.2 gives estimates for potential world-
wide inputs to the oceans. One factor of immediate
concern arising from these values is that in many
cases, and assuming the validity of these numbers,
considerably larger quantities of metals are transported to the oceans by atmospheric pathways rather
than *via* the freshwater input. This is of considerable importance, because metals carried to the oceans
by rivers tend to be held in near-shore and estuarine
"sinks," whereas atmospheric dispersal permits distribution to the surface layers of open ocean areas.
A clear understanding of the characteristics of these
pathways is also required to aid in understanding
terrestrial distributions, since the major effect
on food resources may be from direct particulate
fallout[3] or atmospheric washout rather than from
aqueous uptake.

The principles governing the theoretical distributions of metal species in aqueous systems have
been covered in a general fashion in Chapter 1.

Table 10.2

World Production (1968) of Selected Pollutant Metals
and Potential Inputs to the Oceans (Million Tons/Year)
(Data from Reference 4)

Element	World Production	River Transport	Atmospheric Washout
Ag	0.01	0.01	
As	0.06	0.07	
Cd	0.01	5×10^{-4}	0.010
Hg	0.009	0.003	0.080
Pb	3.000	0.100	0.300
Se	0.002	0.007	
Zn	5.000	0.700	

From known thermodynamic data it is possible in most cases to make reasonable predictions regarding the compositions of coexisting phases under given simplified conditions. One commonly utilized format is that of Eh-pH diagrams, and examples of these for mercury and lead are given below (Figures 10.1 and 10.6). Computed plots such as these must be used with the greatest caution. Apart from the given limitations (such as the necessity of limiting the number of components considered) many of the required constants are not well known; in addition, the relationships imply thermodynamic stability even though unpredicted metastable phases frequently exist under natural conditions. Methyl mercury appears to be one such example.

It is of paramount importance, however, to appreciate—as has been frequently emphasized in these pages—that the trace-metal budgets of all natural waters are largely controlled by physico-chemical uptake to the inorganic and organic particulate phases. Table 10.3 gives log solubility products for some of the metals considered here. Theoretically, the presence of the common anions listed in this table should control the concentration of the metals in the various waters (sulfides will only be controlling under rare localized conditions, of course); this concept is implicit in the practical use of pH/Eh field diagrams. Table

Table 10.3

Approximate Log Solubility Products

	Chloride	*Sulfate*	*Carbonate*	*(Hydr)oxide*[a]	*Sulfide*
Ag	-10			-8	-51
Cd			-14		-28
Hg (I)			-16	-24	-47
Pb (II)		-8	-13		-29
Zn			-11		-24

[a] pH dependent; see Figure 1.1.

10.4 shows--for silver, cadmium, mercury and lead--
the best current values for both the expected and
observed dissolved concentrations in seawater,
this medium being chosen in preference to freshwater because of the more widespread applicability
of the example.   Clearly, marine concentrations

Table 10.4

Marine Distribution of Pollutant Metals

Element[a]	*Soluble Complex*[b]	*Solid Phase*[c]	*Dissolved Conc. (log M)*		*Biota Conc. (log factor)*[fg]	
			Calc.[dg]	*Observ.*[e]	*Plants*	*Animals*
Ag	$AgCl_2^-$	AgCl	-4.2	-8.5	+3.0	+3.0
Cd	$CdCl_2^o$	$CdCO_3$	-4.4	-9.0		
Hg	$HgCl_4^-$	HgO	+1.9	-9.1		
	$HgCl_2^o$		-0.3			
Pb	$PbCl_3^-$	$PbCO_3$	-5.6	-9.8	+4.0	+4.0
	$PbCl^+$		-5.8			

[a] Examples of metals which occur in seawater as chloride complexes
[b] See Table 1.2
[c] Possible controlling least soluble compound; sulfides excluded.
[d] Concentration expected on basis of stated dissolved and solid species.
[e] Data from Reference 5.
[f] Expressed as mean uptake or concentration factor assuming applicability of dissolved values.
[g] Data from References 6 and 7.

are far lower than predicted, based on this particular choice of soluble species and least-soluble compounds. It could be argued that the latter choices are, in fact, poor; however, though certain refinements have been made in this respect since this exercise was initially attempted,[8] the basic conclusions remain unchanged. Table 10.4 also includes marine biological concentration factors, but undoubtedly the most important removal mechanism in this environment is to the sediments in near-shore areas.

In some waters, heavy-metal solubilities are higher than expected from solubility product computations, and one important reason for this is that *all* the contributing species ought to be taken into account but seldom are. As just one example, it has been shown[9] that the solubility of silver in sulfide solution is around 10^{-8} M, whereas the log solubility product of Ag_2S is approximately -50. The solubility in this simple case is determined by the presence of complexes such as $Ag(SH_2)^-$; in natural waters innumerable complexing ligands--especially organic-- should be taken into account. Figure 1.1 shows solubilities of metals in equilibrium with (hydr)oxides calculated from the apparent solubility products but, again, the presence of other soluble species will drastically change this picture.

In recent years it has been observed that reported mean values for the trace heavy-metal contents of natural waters have decreased somewhat, and this is probably a function of improved analytical technique. However, there are still far too few data available even to begin to define natural background levels in any given geochemical environment, and most values are for waters potentially biased by man. Table 10.5 is offered as one example of an integrated study of several bodies of water considered to be uncontaminated.

MERCURY

There is more current concern over the potential deleterious effects of mercury discharged into the environment than over any of the other heavy metals. Although to a large extent the preoccupation with this metal is unjustified, it may be useful to summarize, by way of example, what is presently known about the distribution of this metal in the aqueous environment in more detail than for the other metals considered in this chapter.

Table 10.5

Cd, Cu, Pb and Zn Values for Uncontaminated
Natural Waters[a]   (μg/ℓ)

	Cd	Cu	Pb	Zn
Trinity River Shasta-Trinity National Forest, California	0.2	2.2	2.5	0.3
Sulphur Springs Kittitas County, Washington			1.8	3.2
Park Lake Kittitas County, Washington	1.0	0.2	1.0	4.3
Rachel Lake Kittitas County, Washington	0.2	0.6	<0.2	0.4
Roosevelt Lake Okanogan County, Washington	0.3	1.3	1.6	1.6

[a]Single study using anodic stripping voltammetry.   Data from
Reference 10.   Samples from high altitude, uncontaminated
waters.

   The primary natural sources of mercury are the
native metal and the sulfide HgS. The latter is the
principal utilized source, mined in very restricted
areas of anomalously high concentrations. From
evidence of recent increases of mercury in Greenland
snows,[11] it would appear that global concentrations
are currently increasing. The primary pathways must,
however, be atmospheric rather than aqueous. It has
been suggested[4] that the accelerated releases are
attributable to an increase--albeit man-induced--in
the earth's degassing mechanisms, rather than to
direct industrial discharge. The latter may be
highly significant in localized areas, however, and
wastes from the primary industries that use mercury
(notably chloralkali) are often higher than the
usual level of industrial wastage.[12] Elemental
mercury and a variety of inorganic and organic
complexes are released from the various manufacturing
and agricultural uses. The principal industrial
sources have been noted in Table 2.3 and estimates
of world primary production and recent consumption
in the United States are listed in Table 10.2 and
10.1. Under natural conditions (and considering

only low concentrations of total sulfur and chloride)
the primary stable mercury solid phases are HgO,
Hg_2Cl_2 and HgS[13] coexisting with liquid mercury.
Atmospheric oxidation should convert HgS to free
mercury or to Hg^{2+} but the reaction is very slow
under ambient conditions.

Mercury in the Aqueous Environment

Figure 10.1[13] shows the equilibrium stability
fields of soluble mercury species under normal
atmospheric conditions and with the given anion
concentrations. Clearly, the solubility of elemental
mercury (around 25 µg/l) should control the natural
levels of mercury found in many waters. Under mild
reducing conditions, mercury will be removed from
the water by precipitation as HgS but may also rarely
be remobilized if a very strong reducing environment
occurs. With an increase in chloride concentration,
the $HgCl_2^\circ$ field of Figure 10.1 initially expands
and, in the presence of marine halide concentrations,
the complex chloride species given in Tables 1.2 and
10.4 are stable. This simplified representation is
ideal under the stated constraints but does not very
accurately reflect conditions likely to be encountered
in fact. In the first place, the usual limitations
of such representations apply; reaction rates of
thermodynamically favorable reactions may be very
slow under natural conditions, and the many other
possible complexing inorganic and organic ligands
have not been considered. Thus methyl mercury is
not stable relative to the other compounds of Figure
10.1, yet it is known to occur widely in nature;
mercury would also be expected to form natural
chelate complexes. Secondly, the aqueous transportation pathways will almost certainly be dominated--
as with all heavy metals--by other physical and
biological processes of the types noted above.
Figure 10.2 is an attempt to show the major
transport pathways of mercury in the natural environment. Some routes may be considered as cyclic
in the conventional manner, but, in terms of essentially short-term aqueous pollutional stresses, the
various "sinks" are of controlling importance;
knowledge of rates and mechanisms of possible
remobilization from these phases is an urgent
necessity. The latter is a very weak link in our
current understanding of the environmental mobility
of heavy metals. Mercury methylation is one of the
few reasonably well documented[14,15] phenomena of
this type.

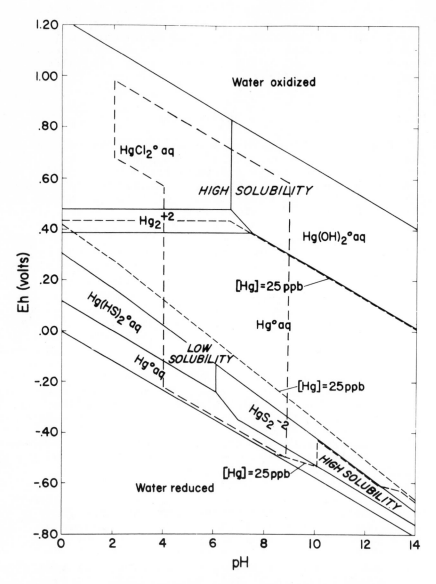

Figure 10.1.   *Eh–pH stability fields for aqueous mercury*
species in solution containing 1mM chloride and
sulfate at 25° C and 1 atmos.   Natural water
limits of Figure 1.3 superimposed.   (Reprinted
from Reference 13 by courtesy of author.)

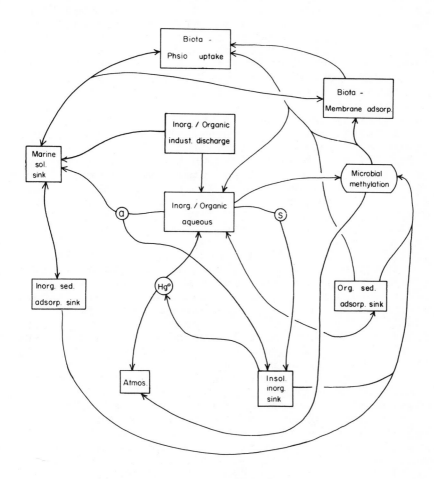

Figure 10.2. Partial and simplified natural aqueous mercury pathways.

^{a}*Adsorption processes.*

Jenne[16] has well reviewed the status of work on the sorption removal of aqueous mercury species. As with most other heavy metals (but possibly not cadmium), mercury appears to be strongly sorbed by contacting sediments, and this scavenging mechanism is probably chiefly responsible for maintaining low natural soluble levels of these metals. It has been noted elsewhere that solute species of mercury in

seawater rapidly decrease away from point injection
sources. At least one similar freshwater reduction
from a discharge anomaly has also been recorded,[17]
and comprehensive studies of mercury distribution
within one unified river system[18] and a lake[19] have
further demonstrated the importance of uptake on
sediments. The converse remobilization processes--
by microbial activity, chemical reaction or incorporation into various organisms--are considerably
less clearly understood or appreciated.

Concentration of mercury over coexisting aqueous
contents by various members of the freshwater and
marine biosphere is well known and has been the
principal source of the pollution controversy.[20]
In this respect, of course, mercury is behaving no
differently than all the other heavy metals, except
that certain biota species have a propensity for
enriching particular elements. Mercury appears to
be particularly high in certain of the larger fish
species (tuna, swordfish) and in marine mammals,
for example, but this may largely reflect collection
biases. Such bioconcentrations constitute an important mechanism[21] for marine transport and distribution
of the metals away from near-shore areas, but it is
highly likely that these contents are predominantly
"natural";[22] there is evidence of anthropogenic stress
in remarkably few cases and no food-chain amplifica-
tions have been positively identified.

The uptake by members of the aqueous biological
community initially may be "active"--a direct
physiological incorporation--or may be a "passive"
surface sorption. It would appear, however, that
major fish mercury toxicity is due to the latter
process, specifically to adsorption and reaction
with the gill epithelium. This is believed to be
so also for cadmium, copper and lead poisoning, but
apparently not for zinc.[23,24] Sorption may be
followed by further incorporation and transportation
within the organism. The final concentration sites
are the major body organs. We have found, for
example (Table 10.6), the highest levels of mercury
in the kidneys and livers of salmon species and
negligible amounts in the eggs. However, it is
important to note that mercury accumulations have
been shown to be "self-flushing," and also that gill
"poisoning" may lead to sloughing off of metal-rich
mucus;[25] hence the rates of metal build-up and
contact times of the organisms are very important.
It might be reasonably argued, in fact, that fish
toxicity data of the type reproduced in Table 10.7
are essentially useless--misleading at best.

Table 10.6

Mercury Contents of Mature Sockeye Salmon Taken from
The Gulkana River, South Central Alaska
Mean of Duplicate Determinations (μg/g dry weight)

Sample	Sex	Muscle	Liver	Kidney	Eggs
1	m	0.05	0.16	0.26	
2	f	0.07	0.22	0.25	0.01
3	m	0.03	0.21	0.20	
4	f	0.05	0.12	0.18	0.01
5	m	0.03	0.20	0.13	

Table 10.7

Lethal Concentrations (μg/ℓ) of Various Mercury Compounds
for Some Fish Species
(From Reference 26; see text for utility of these data)

Fish	Compound	Lethal Conc. (μg/ℓ)
Stickleback	Mercuric nitrate	20
	Mercuric chloride	20
Guppy	Mercuric nitrate	20
	Mercuric chloride	20
Shiner	Ethyl mercury phosphate	800
Eel	Mercuric chloride	27
Catfish	Phenyl mercuric acetate	580
	Ethyl mercury phosphate	1,300
Rainbow trout	Pyridyl mercuric acetate	2,000
	Mercuric chloride	9,200
Salmon	Phenyl mercuric acetate	20
	Mercuric acetate	50

Considerably less is known about the influence of
mercury on the lower trophic levels. Although
there has been some experimental work on inhibition
of phytoplankton photosynthesis,[27] no natural
perturbations have been recorded.

Atomic Spectrometric Analysis

Conventional flame atomization AA procedures
for mercury suffer from poor detection limits and
sensitivity (see Tables 3.5 and 6.1); analysis by
AF is somewhat better (Table 3.6). Some success
has been reported[28] for vacuum U.V. analysis but
this wavelength region is inaccessible to standard
equipment. By far the largest number of environmental mercury data are now obtained by flameless
AA analysis, utilizing the appreciable atomization
of this metal at ambient temperatures. The principles of this cold-vapor method have been long
appreciated, but practical AA applications whereby
the sample solution is chemically reduced to evolve
metallic mercury vapor which is subsequently flushed
through a long path-length, enclosed glass cell
positioned within the spectrometer source beam were
only fairly recently popularized.[29,30] Unlike the
high-temperature flameless atomization devices
discussed earlier, the mercury vapor is flushed
through the cell (using some inert carrier) over
an extended period, and the recorder signal may require several minutes to reach a maximum and decay
(Figure 10.3) so that the signal trace theoretically
should be integrated. It is important to appreciate
here that severe losses of the mercury can occur
within the apparatus between the generation and
atomization modules. For example, Fritze and Stuart[31]
have noted that a 2-3% loss of mercury vapor may
occur during the addition of the reducing reagents,
and also that up to 20% of the mercury may be adsorbed
within the drying column. These authors have suggested, in fact, that water vapor may not be a
potential interferent within the atomization cell
as is commonly supposed, but that a carry-over of
aerosol may cause problems. Subsequent published
developments of this technique have been principally
refinements of this basic approach, and commercial
analytical modules are available which are specific
for mercury by this method. One such instrument,
the Mercury Monitor (Laboratory Data Control Inc.)
incorporates extra long (and dual) flow cells and has
a claimed detection limit of 0.1 ng, sufficient for

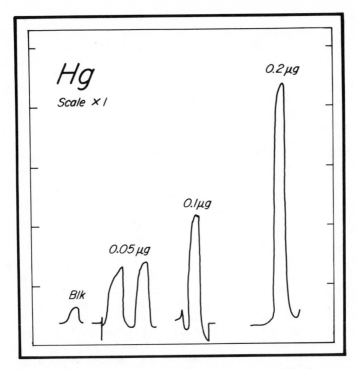

Figure 10.3. Recorder traces for mercury standards by flameless AA analysis.

the analysis of soluble mercury in many natural waters. Other workers[32] have also claimed direct natural water detection capabilities for various instrumental arrangements but, in general, a pre-reduction concentration of the water sample will be required in the usual fashion. A dithizone extraction is most often used for marine water samples and also for some freshwater applications,[33] but for the latter samples, a considerably more elegant approach involves a pre-analysis collection of the evolved metallic mercury. Kalb[34] and Fishman[35] advocate amalgamation of the atomic mercury with silver followed by a rapid remobilization (in the absence of water vapor or other potential inter-ferents) by, respectively, induction furnace or direct electrical heating of the silver. By these means, not only is a concentration of the initial sample effected, but the recorded absorption peak is rapidly developed. Another variation of this technique utilizes direct electrolysis of the

elemental mercury onto some suitable cathodic
material (usually copper) in place of the amalgamation step, with the same rapid remobilization by
subsequently heating the electrode.[36-38] There is,
however, some question as to whether this electrolytic process is applicable to highly saline samples.
The indigenous mercury contents of the aqueous
biota are considerably higher than the mercury
content of the waters so that, for the analysis of
these materials, the principal problem is not
detection limitation but treatment of the sample
for a total, reproducible release of the mercury
(see also Chapter 5). Various digestion-oxidation
steps have been proposed, and sulfuric acid or
sulfuric-nitric treatment plus a permanganate
oxidation[39-41] are typical of these. Subsequent
reduction of the mercury compound to the metal may
be effected by stannous chloride with or without
hydroxylamine. In an interlaboratory comparative
study[42] of mercury determinations from fish, an
overall recovery value of 83-95% was given by a
standardized method very closely akin to those
referenced above. This latter study, however,
points up the possibility of differential release
of inorganic and organic bound mercury. Since most
biotic mercury is believed to be methyl mercury,
formulation of a correct dissolution procedure is
of considerable importance. The comparative study
cited above specifically included samples spiked
with methyl mercury, and better than average recovery of the mercury vapor was reported for these
samples. However, it has been suggested[43] that
inorganic mercury only is released by stannous
chloride reduction. This latter author thus sug-
gests a chemical species differentiation based on
a controlled release of organic bound mercury using
a mixed stannous chloride-cadmium chloride-sodium
hydroxide reagent. No major problems associated
with the analysis of mercury from inorganic particulate material have been reported.[29,44] Joensuu[45]
has described a total detection system for the direct
release and determination of mercury from sediments.
The detection limit for mercury by flame-cell
AF analysis (Table 3.6) is considerably better than
that by continuous AA nebulization. Winefordner
and co-workers[46,47] for example, have published
early work based on high temperature oxyhydrogen
and oxyacetylene flames; the use of electrodeless
discharge sources has given[48] a detection limit of
80 μg/l. However, only AF techniques based on the
reduction--cold-vapor aeration atomization discussed

above are likely to be applicable to natural water
analysis.[49],[50] April and Hume[51] have described an
emission analysis method, using a low pressure
helium plasma source, that has a claimed detection
limit of 2 ng of mercury and essentially the same
system has been used as a gas chromatographic detector for the determination of methyl mercuric
compounds in fish.[52] Cook *et al.*[53] have noted that
both AF and AE mercury analysis might benefit by the
addition of a photon counting detection system. The
accelerating interest in monitoring environmental
mercury levels has spawned a number of specific
"mercury meter" instruments based on the cold-vapor
aeration procedure. The portable photometer of
Ling[54] should be noted, and Bailey and Lo[55] have
described an automated "Hatch and Ott" system.

Other Analytical Techniques

Table 10.8 is intended to summarize the characteristics (of particular relevance to the water
chemist) of the commonest analytical techniques
applicable to mercury. Arc spectrography has been
applied to biota samples,[56] but neither this technique nor X-ray fluorescence are likely to be much
used in the average environmental monitoring
laboratory. Colorimetry is, and will continue to
be, widely utilized for water analysis (Table 2.7)
while it remains a designated standard method
(Chapter 2). Dithizone extractions yield reasonably
precise analyses with low detection limits (Table
2.13), but extra care must be exercised where
appreciable quantities of organic matter occur.

The other major analytical techniques comparable
with AS for water analysis are those of polarography
and thermal neutron activation. The latter is a
sensitive, total analysis (speciation may be accomplished by pre-analysis separations as for atomic
spectrometry), but it is expensive and decidedly
nonportable. Pre-irradiation concentration is
inevitably required for natural water samples and
many possible techniques for this are available.
Only two such examples--sulfide coprecipitation,[57]
and use of anion exchange resins[58]--will be
referenced here. Anodic stripping voltammetry from
graphite electrodes is potentially applicable to
most natural water samples (organic complexes must
be pre-oxidized), but a comprehensive study has yet
to be published.

Table 10.8

Application of Various Techniques to Mercury Analysis

	Spectrog-raphy	X-ray Fluores.	Colorim-etry	Neutron Activation	Polarog-raphy	Atomic Spectromet.
mg/ℓ Range	X	X	X	X	X	X
μg/ℓ Range						
Direct Solid Analysis	X	X		X		
Solution Analysis	X		X	X	X	(X)
Mandatory Pre/Post Analysis Chemistry			X	X		X
Potential Direct Analysis (μg/ℓ Range)					(X)	
Potential Automation			X		(X)	X
Total Analysis Only	X	X	(X)	X		X
Potential Soluble Speciation			(X)		X	

There are innumerable procedures in print concerned with the determination of mercury in aqueous biota. For analysis by neutron activation,[59-62] the [197]Hg γ photopeak (77.3 KeV; 65 hr) is usually counted, and various radiochemical separations employed. The principal problem, however, is not associated with the final analysis but with the initial digestion and release of the metal from the organic complexes, and this problem--being common to all biota analytical methods--cannot be overemphasized; the covalent bond between mercury and carbon is, in fact, remarkably inert and chemically inactive. Several common digestion procedures have been referred to above and the literature is replete with others, but the analyst would be well advised to perform routine recovery checks on whichever procedure is selected. Westöö[63,64] has given a combined thin layer and gas chromatographic method for the direct determination of biotic methyl mercury. This work utilized an easily poisoned electron-capture detector; an alternative plasma emission detector system has been cited above.

SILVER

Very little is known of the natural distribution patterns of silver. The soluble content of freshwaters may be controlled by the fairly ubiquitous occurrence of chloride ions, and the metal should be virtually absent in sulfide-bearing waters. In highly saline waters, a series of soluble chloride species should exist as given in Tables 1.2 and 10.4 but, as noted previously, controlling mechanisms other than chloride equilibria are usually operative. In the studies of Kharkar *et al.*[65] silver was adsorbed onto various particulate phases least of all the metals tested--a surprising result in view of the troublesome uptake of this metal onto container surfaces noted below. Table 10.4 shows a concentration factor for marine organisms over seawater of 10^3; factors in the range 10^4-10^5 have been recently suggested.[66] Atmospheric particulate transportation is probably an important pathway to the oceans, and urban air enhancements have been recorded.[67]

Atomic Spectrometric Analysis

Silver may be determined to very low detection
limits by AS analysis (Tables 3.5, 3.6, 7.5-7.10),
but several severe practical problems--associated
with precipitation and sorption onto containing
surfaces from solution--must be guarded against.
The principal problem is that of halide formation.
Halides dissociate readily in flames (see β data of
Table 7.1) and therefore atomization poses no
problems--even for colloids; however, these compounds
precipitate (and adsorb) on standing, and, since
halides are ubiquitous in nature, problems abound
for the environmental chemist. It has been noted[68]
that 50% hydrochloric acid will hold silver in
solution, but it is not recommended that an acid of
this strength be nebulized in the average AS equipment. Addition of diethylene triamine is also held
to prevent precipitation (see also Reference 69).
The very pronounced sorption of silver compounds
onto samples and other container surfaces has long
been regarded as a major potential source of error
in characterizing aqueous silver distributions.
Adsorption effects on various surface materials
have been documented by West and co-workers,[70,71]
and these people have advocated the use of a thio-
sulfate additive to stabilize silver in solution
over short periods. A correlation between pH and
the degree of adsorption has been recorded[72] and
Dyck[73] has shown that lowering the solution pH to
around unity drastically reduces the sorption onto
both glass and PVC surfaces. The silver may also
be complexed with any of several organic ligands
(EDTA, for example; see Reference 74), but the
analyst must be well aware of potential effects upon
the subsequent processing; thus, if a solvent extraction concentration step is to be used, the chelate
must be extractable into the solvent at that pH
range. These pre-analysis silver precipitation
effects may occur also at the atomization stage,
and the absorption signal for silver in conventional
flames has been noted[75] to be a function of the
particular chelate complex aspirated. This latter
phenomenon--a typical condensed phase interference--
is common enough in AA analysis but seldom fully
appreciated.
Flame AA determination of silver in natural
waters has long relied on a pre-analysis concentration
step in the usual fashion. There should be no particular problems associated with these manipulations
so long as the history of the sample is well understood,

as discussed above. Chao[76] has advocated a two-stage
concentration procedure using an anion exchange resin
(Table 5.3) to remove silver from acidified samples
followed by an APDC/MIBK extraction, or TOTP into
MIBK alone. TOTP is a most useful chelating agent[77]
and is selective for silver.

Silver is also very efficiently determined by
flame AF analysis (Table 3.6). West and Williams,[78]
for example, have used a standard high-intensity HC
lamp as a source and have determined silver with
good precision to the 10 µg/l range; this may be
bettered by prefacing with a concentration step (the
latter cited authors extracted the metal as the
di-n-butylammonium salicylate).

Silver was one of the first metals found to be
amenable to nonflame AS analysis. Chao and Ball[79]
used the tantalum sampling boat described in Chapter
7. Published detection limits for silver by discrete
filament and furnace AA and AF analysis (Tables 7.8
and 7.9) suggest the possibility of a direct analysis
from many natural waters, but, to date, no data thus
obtained have been published.

Other Analytical Techniques

Many other analytical techniques are in use for
the determination of trace amounts of silver in a
range of materials (see the review by Beamish *et
al.*,[80] for example), but, for natural water samples,
only colorimetry[81] and neutron activation (with
radiochemical separation; *e.g.*, Reference 65) are
considered suitable unless the metal is present in
unusually high concentrations. In the foreseeable
future, nonflame AS methods should have no serious
rivals for aqueous samples.

ARSENIC AND SELENIUM

The physiological cycles of both these elements
are reasonably well understood and there has been
considerable work on soil-plant-animal pathways.
There is, however, a paucity of data describing
environmental levels in natural water systems. The
literature is replete with statements such as:
"Relatively few arsenic determinations have been
made in natural waters ... because of analytical
difficulties";[82] and "Little is known about the
role selenium plays in aquatic ecosystems."[83]

Selenium is a nutrient at low concentrations
but toxic to most organisms in the mg/l range. The
presence of this metal in natural waters may be due
to leaching from selenium-rich soils or to commercial wastes from, for example, the electronics
industries, but atmospheric pathways probably
predominate so that selenium may potentially be
globally dispersed. Hashimoto *et al.*[84] have studied
atmospheric selenium data in some detail and have
noted a parallel distribution trend with sulfur
derived from fuel materials. Easily detectable
quantities have been shown to occur in both freshwater and saltwater fish[85] and in the organs of
polar marine mammals, but at the present time not
enough is known of the baseline concentrations in
these systems to allow speculation concerning
possible pollution anomalies.

Much the same situation holds with regard to
the environmental distribution of arsenic. Atmospheric dispersal probably predominates, but too few
aqueous data are available to permit more than a
cursory evaluation of the pathways. Where point-
sources of pollution (discharging directly into
aqueous systems) have been monitored, the situation
appears to be closely akin to that for mercury and
cadmium; that is, the metal is rapidly removed from
the water column to the sediment "sink," and nothing
definitive is known of possible remobilization
mechanisms although "high" levels have been reported
from some benthic organisms. A possible pollutional
source of arsenic from household detergents has been
reported.[86] Severe poisoning of man can arise from
ingestion of as little as 10 mg of arsenic, and
carcinogenic properties are imputed.

Atomic Spectrometric Analysis

There are difficulties with the determination
of both arsenic and selenium by continuous flame AA
analysis because the major resonance lines lie in
the far UV region, and below around 2000 Å the
opaqueness of the flame gases seriously diminishes
the analytical sensitivity. It has been noted[87]
that entrained-air flames offer advantages for
absorption measurements in this spectral region,
and Amos *et al.*[88] have proposed use of an entrained-
air flame coupled to a long path-length absorption
tube. Flame AA is, however, insufficiently sensitive
to permit the direct determination of either arsenic
or selenium at the concentration levels usually

found in natural waters, so that preconcentration procedures of the general type discussed in Chapter 5 have been required. Detection limits of 2 and 20 µg/ℓ for selenium and arsenic respectively have been published[89] using an APDC-carbon tetrachloride solvent extraction with an additional aqueous back-extraction step (a total concentration of x 100), and coprecipitation has also been advocated.[90] Fortunately, both these metals yield volatile hydrides[91] which may be utilized both to preconcentrate the metals where necessary and as the basis of enclosed atomization mode analysis as considered in Chapters 5 and 7. The arsine or hydrogen selenide may be introduced as a discrete "sample" into a hydrogen-argon-entrained air flame, but decomposition within an electrically heated tube atomizer[92] is a more elegant approach which yields sub-µg/ℓ detection limits. Goulden and Brooksbank[93] have subsequently improved this procedure by flushing into the tube using an argon-air mix. The required hydrogen is generated during the reduction step. The hydrides may be concentrated if necessary prior to atomization by initial collection in a nitrogen cold-trap.[94] These procedures are eminently suitable for the analysis of biota also in a fashion analogous to that previously noted for mercury, although it appears that residual traces of perchloric acid can severely suppress the selenium signal. Figure 10.4[95] illustrates a simple apparatus for the hydride generation. One other indirect AA procedure for arsenic has been proposed by Danchik and Boltz,[96] and Ihnat,[97] by first separating the element from the matrix, has determined the selenium content of biological tissue *via* standard carbon filament atomization.

Flame AF and AE analysis procedures for arsenic and selenium generally give mg/l range detection limits and are hence only applicable to the particulate phases,[98],[99] but Lichte and Skogerboe[100] have coupled the arsine generation technique to plasma emission analysis. Dean and Fues[101] have outlined a specific AE procedure for arsenic in organic compounds which involves dissolution into an organic solvent and aspiration into a fuel-rich oxyacetylene flame.

Other Analytical Techniques

The standard methods are both colorimetric: that for arsenic depends upon a Marsh reaction

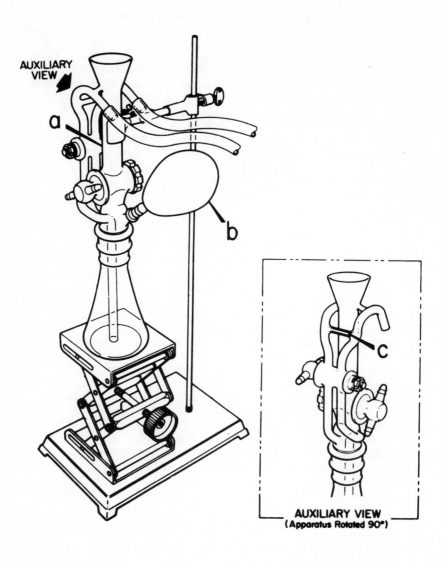

Figure 10.4.   One postulated closed system for the generation
of arsine and hydrogen selenide for subsequent
trace AA analysis of these metals.   (Reprinted
from Reference 95 by courtesy of the author.)
[a]Dosing column.
[b]Balloon reservoir.
[c]Inert-gas bypass.

evolution of arsine and subsequent reaction with
Ag-DEDTC in pyridine; the selenium reaction is with
diaminobenzidine (Table 2.7). Arsenate in water
may also be measured as part of the phosphate
determination using the standard molybdate complex.[82]
Portman and Riley[102] have suggested a photometric
procedure for arsenic in seawater which incorporates
a preliminary thionalide coprecipitation (see
Chapter 5).

Neutron activation analysis is applicable to
both arsenic and selenium in natural waters but a
preconcentration is usually required to achieve the
necessary sensitivity. It should be noted here
that, although major error incurred during radiochemical separations are determinable, preirradiation
manipulation of the samples is to be avoided wherever
possible. Unfortunately, the separation treatments
frequently necessary prior to activation are considerably more complex, and hence more prone to
error, than those commonly utilized for AS analysis.
Ray and Johnson[103] advocate the thionalide coprecipitation noted above. However, selenium is
determinable in nonmarine waters using only post-
irradiation radiochemical separations (*e.g.*, References 11 and 65). Blotcky *et al.*[104] have discussed
the optimum procedures necessary for determining
selenium in biological material *via* neutron
activation analysis.

Myers and Osteryoung[105] have given a preliminary
evaluation of the feasibility of using differential
pulse polarography to determine trace level arsenic,
and several methods for determining soluble selenium
by standard polarographic techniques are available.[106]
Selenium may also be analyzed by fluorometry (*e.g.*,
Reference 107 and--for sediments--Reference 83) and
by gas-liquid partition chromatography.[108]

CADMIUM

Cadmium is extremely toxic and has a long biological half-life in man. As with the majority of
pollutant elements considered in this chapter, the
major transportation pathways appear to be atmospheric, and concentrations reported for urban air[67]
are high in relation to most other heavy metals
monitored. Human uptake mechanisms may give rise
to markedly different concentrations; approximately
40% of inhaled cadmium is absorbed by the body while
the equivalent figure for the metal ingested is only

around 5%. An excellent review of the effects of cadmium on man has been given by Friberg *et al.*[109]
    The current (1971) U.S. consumption of this metal is in excess of 15 million pounds (with the electroplating industry as the primary consumer) but, as with all the metals discussed here, natural aqueous base levels and distributions are very poorly understood. The major documented cases of obviously anomalously high levels of cadmium in natural waters have been very localized "hot spots" associated, for example, with zinc mining wastes.[110] Abdullah *et al.*[111] have demonstrated that concentrations up to 20 to 40 times open sea values occur (in coastal waters in western areas of the United Kingdom) adjacent to point-sources of both mineralized and industrial discharge. Such very high concentrations appear to be quite localized, as is the case also with, for example, soluble arsenic and mercury, and it might be supposed that the behavior of cadmium would closely parallel that of zinc in general and that rapid removal from the hydrosphere should be the norm. There is some evidence, however, that cadmium may be considerably less susceptible to abiotic removal than most of the other heavy metals studied in any detail to date. Preston *et al.*,[66] for example, have shown that around 80% of marine cadmium is not removed by a 0.22 μ membrane filter, whereas the equivalent fraction for mercury[112] falls in the range 10-20%. Inshore levels of soluble cadmium may be a factor of ten higher than the open ocean range of 0.01-0.05 μg/l.[67]
    The cadmium marine biota concentration factor is commonly 100-1000, and large concentrations have been found in the internal organs of fish, shellfish and marine mammals--including up to 500 μg/g in certain Californian sea otter kidneys.[67] The latter occurrence suggests a possible food-chain link from shellfish which is of considerable interest since it would provide evidence for the amplification of at least one heavy metal through an aqueous food web.

Analytical Techniques

    There are several excellent techniques available for the determination of cadmium present in trace quantities in environmental materials and the analyst should experience little difficulty selecting one suited to his particular needs. The detection limits for both cadmium and zinc by AS analysis are

equal to the best that these techniques can offer.
Figure 10.5 illustrates the AA determination of
cadmium in a marine sample by a standard addition
technique. Standard aliquots have been added to

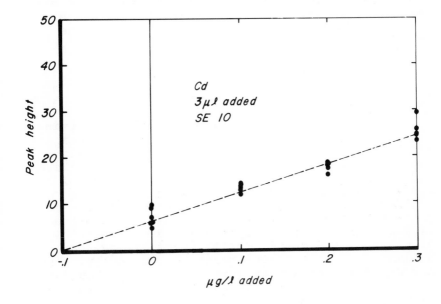

Figure 10.5. *Standard addition AA determination of soluble*
cadmium in a seawater sample via a dithizone
extraction (and X10 concentration) into
chloroform and atomization from a West-type
carbon filament.

replicate subsamples prior to chelation and extraction (with x 10 concentration) into chloroform. The
latter extracts have been atomized from a West-type
carbon filament (see also Figure 7.17). One published method[113] for absorption tube AA analysis of
biological material stresses the effects of sodium
interference, and this is a potential problem which
the analyst must be completely aware of. Many
studies of the optimum determination of cadmium by
AF analysis are available[46],[47],[114] including
information on the best use of EDT sources.[115,116]
    Neutron activation analysis is not a suitable
technique for natural waters unless a large

pre-analysis concentration is deemed acceptable, but this technique[117] has been frequently applied to biota. Trace level cadmium may be determined by a variety of polarographic techniques--it is one of the few metals particularly amenable to the various anodic stripping voltammetry methods--and a plethora of practical applications have been presented in the standard literature.[111] Colorimetry and specific-ion electrode methods[118] offer the potentiality for continuously monitoring some waters.

LEAD

Lead is a widely dispersed naturally occurring element, and one used in a multitude of industrial processes. The major sources of environmental pollution are from gasoline, pesticides, fertilizers and the smelting industry (Table 2.3), but atmospheric lead pollution derives overwhelmingly from the combustion of tetraethyl lead gasoline additives--over a quarter million tons in 1968 alone. These lead products enter the atmosphere as aerosol particles of varying (but mostly small) sizes; it has been estimated[119] that 50-80% range less than 1 μ. The larger particles are deposited (by dry fallout or wet washout) close to the source--adjacent to major highways, for example.[120,121]

Finer particles may travel long distances by atmospheric transport processes prior to removal.[122] Thus, over the last few decades, the lead content of Greenland ice has increased to some 500 times the prehistoric background levels[123] and some surface marine waters have lead contents estimated to be approximately three to six times the prehistoric baseline value.[124] It is believed[4] that the residence time of this oceanic surface anomaly is quite short, of the order of several years only. Terrestrial atmospheric deposition of particulate lead may have a considerable effect on food crops, especially adjacent to urbanized areas. In fact, human contact with this metal is likely to be considerably greater from the food we eat than directly from the air, but a much smaller fraction of ingested lead is absorbed compared with that from respiratory intake.

The distribution of lead (derived from either natural or artificial sources) in natural waters is not well known, but basic thermodynamic data suggest it should be very low. For "normal" waters in

contact with atmospheric carbon dioxide (Figure
10.6) Pb^{2+} may be present where acid conditions
persist. If sulfur species are added to the $Pb-CO_2$
system illustrated, the solute field boundary is
shifted to yet lower pH values (see References 9,
125). However, as in the case of mercury considered

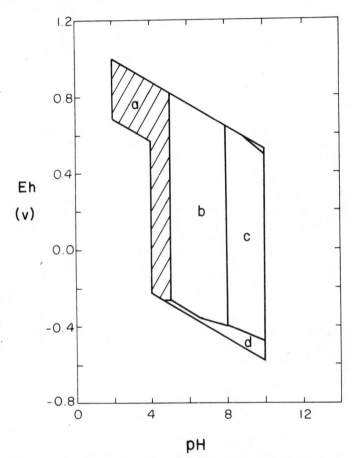

Figure 10.6. *Eh-pH stability relationships for aqueous lead
and carbonate (approx. 10^{-2} M) species only.
(After References 9, 125.)*

[a] Soluble Pb^{2+}.
*[b,c,d] Solid phases: carbonates and elemental
lead. Note that lead is only stable in solution
in this system at low pH values and that with
the addition of other common ligands (sulfur
and chloride species, for example) this field
is reduced further.*

above, soluble concentrations may be higher than
those predicted by such essentially simplistic
inorganic equilibrium models. It has been shown
for example,[126] that the NTA component currently
being substituted for phosphate in detergents may
solubilize the "fixed" lead of sediments. The
major lead contents of the natural water systems
are undoubtedly to be found in the biosphere and
sorbed on inorganic particulates, and the runoff
contribution to the oceanic budget is considerably
less than that derived from the atmosphere.

Atomic Spectrometric Analysis

     In most waters, natural lead concentrations,
and potential interferents are such that a preflame
AS analysis extraction/concentration process is
mandatory, and many standardized techniques (some
directly applicable to biological samples also) are
available in the literature (*e.g.*, References 127,
128). However, at the present time, very many
routine biological analyses (blood and urine, for
example) are being performed using flame-activated
discrete samplers of various designs, since lead
may be appreciably atomized at relatively low
temperatures (see Table 7.1). The open tantalum
"boat" design noted in Chapter 7[129] has been followed
by various types (specifically designed for clinical
lead determinations) which atomize the metal from
a cup device into a containing tube or cell. The
Delves "cup" model[130] (Figure 10.7) has been
widely distributed, and other similar modules
(*e.g.* Figure 10.8[131]) are also available. A flameless continuous monitoring technique for atmospheric
lead has also been recently described.[132] From a
review of applications such as these, it might be
inferred that the more efficient nonflame atomizers--
the furnaces and filaments of Chapter 7--should yield
environmental aqueous lead values in certain cases
by direct analysis. Although no such data are available as yet, our own initial tests confirm that this
should indeed be so in the absence of interfering
matrix effects, and Woodriff and Lech[133] have determined lead in air using furnace atomization.
Several comprehensive published applications of AF
analysis to lead are available[134,135] and also an
emission technique for biological lead contents
using a demountable HC device.[136]

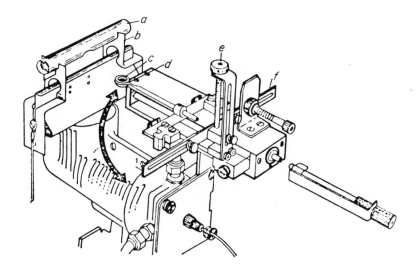

Figure 10.7. *Nickel crucible ("Delves' cup") and absorption*
tube modification of the Perkin-Elmer Corp.
discrete boat[137] atomizing assembly for trace
lead analysis. (Reprinted from Reference 130
by courtesy of the Society for Analytical
Chemistry.)

[a]10-cm nickel absorption tube.
[b]Supports for attaching tube (a) to burner.
[c]Nickel crucible.
[d]Platinum wire holder for crucible (c).
[e]Vertical adjustment screw.
[f]Horizontal adjustment screw.

Other Analytical Techniques

The standard method for aqueous lead is by
dithizone colorimetry (Table 2.7), and this has also
formed the basis for experimental AutoAnalyzer continuous monitoring systems. Specific-ion electrodes
have a realizable utility to 10^{-7} M[138] and several
excellent studies on the use of anodic stripping
voltammetry are available.[10,139] The latter authors
claim a sensitivity of 10^{-9} M lead in natural waters
(see also Table 10.5) and demonstrate further how
shifts in the oxidation potential may indicate
complexation mechanisms.

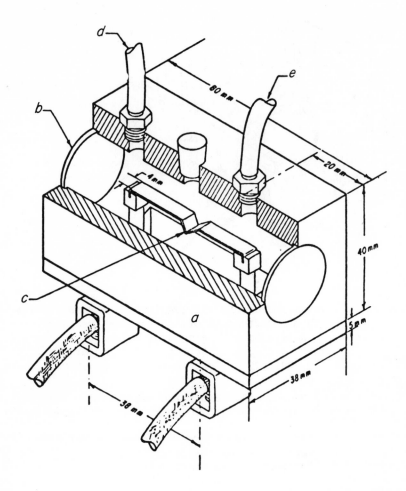

Figure 10.8. Modification of Donega and Burgess[137] absorption tube method for the determination of lead. (Reprinted from Reference 131 by courtesy of the American Chemical Society.)

^a*Absorption tube (Lexan).*
^b*Quartz window.*
^c*Tantalum strip to hold liquid sample.*
^d*Argon inlet.*
^e*Argon outlet.*

Conventional neutron activation analysis makes use of the ^{208}Pb (n, γ) ^{209}Pb reaction, but this has a short half-life (3.3 hours) and is not very practical for most routine uses. High energy γ-photon activation, however, yields nuclides which decay with usable γ-ray emission and this technique has been used[140] for the lead contents of some biological materials.

REFERENCES

1. Barnett, P. R., M. W. Skongstad, and K. J. Miller. *J. Am. Water Works Assoc., 61,* 61 (1969).
2. Knapp, C. E. *Environ. Sci. Technol., 5,* 584 (1971).
3. Hadjimarkos, D. M. *Environ. Sci. Technol., 5,* 1048 (1971).
4. National Academy of Sciences. *Marine Environmental Quality,* Ocean Sci. Comm., National Academy of Sciences, Washington, D.C., 1971.
5. Goldberg, E. D. in *Chemical Oceanography,* Vol. 1 (J. P. Riley and G. Skirrow, eds.), Academic Press, London, 1965, pp. 163-196.
6. National Academy of Sciences. *Radioactivity in the Marine Environment,* Panel on Radioactivity in the Marine Environment, Comm. on Oceanography, National Academy of Sciences, Washington, D.C., 1971.
7. Schindler, P. W. in *Equilibrium Concepts in Natural Water Systems* (R. F. Gould, ed.), Advan. Chem. Ser. No. 67, American Chemical Society, Washington, D.C., 1967, pp. 270-285.
8. Krauskopf, K. B. *Geochim. Cosmochim. Acta, 9,* 1 (1956).
9. Stumm, W. and J. J. Morgan. *Aquatic Chemistry,* John Wiley and Sons, New York, 1970.
10. Zirino, A. and M. L. Healy. *Environ. Sci. Technol., 6,* 243 (1972).
11. Weiss, H., M. Koide, and E. Goldberg. *Science, 175,* 692 (1971).
12. Goldwater, L. J. *Sci. Am., 224,* 15 (1971).
13. Hem, J. D. in *Mercury in the Environment,* Geological Survey Prof. Paper 713, U.S. Gov't. Printing Office, Washington, D.C., 1970, pp. 19-24.
14. Wood, J. M., F. S. Kennedy, and C. G. Rosen. *Nature, 220,* 163 (1968).
15. Jernelov, A. *Limnol. Oceanog., 15,* 958 (1970).
16. Jenne, E. A. in *Mercury in the Environment,* Geological Survey Prof. Paper 713, U.S. Gov't. Printing Office, Washington, D.C., 1970, pp. 40-45.

17. Dall'Aglio, M. in *Origin and Distribution of the Elements,* (L. H. Ahrens, ed.), Pergamon Press, New York, 1968, pp. 1065-1081.
18. Cranston, R. E. and D. E. Buckley. *Environ. Sci. Technol., 6,* 274 (1972).
19. Armstrong, F. A. J. and A. L. Hamilton. ACS Symposium on Metal-Organic Interactions in Natural Waters, April 1972, Boston, Mass.
20. United Kingdom Ministry of Agriculture, Fisheries and Food. *Survey of Mercury in Food,* H. M. Stationery Office, London, 1971.
21. Klein, D. H. and E. D. Goldberg. *Environ. Sci. Technol., 4,* 765 (1970).
22. Miller, G., P. Grant, R. Kishore, F. Steinkruger, F. Rowland, and V. Guinn. *Science, 175,* 1121 (1972).
23. Lloyd, R. Proc. 3rd Sem. on Biological Problems in Water Pollution, U.S. Dept. of Health, Education and Welfare, Washington, D.C., 1962.
24. Wojtalik, T. A. *J. Water Pollution Control, 43,* 1280 (1971).
25. McKone, C. E., R. G. Young, C. A. Bache, and D. J. Lisk. *Environ. Sci. Technol., 5,* 1138 (1971).
26. Greeson, P. E. in *Mercury in the Environment,* U.S. Geol. Surv. Prof. Paper 713, U.S. Gov't. Printing Office, Washington, D.C., 1970, pp. 32-34.
27. Harriss, R. C., D. B. White, and R. B. MacFarlane. *Science, 170,* 736 (1970).
28. L'vov, B. V. in *Atomic Absorption Spectroscopy,* (R. M. Dagnall and G. F. Kirkbright, eds.), Butterworths, London, 1970, pp. 11-34.
29. Hatch, W. R. and W. L. Ott. *Anal. Chem., 40,* 2085 (1968).
30. Manning, D. C. *Atomic Absorption Newsletter, 9,* 97 (1970).
31. Fritze, K. and C. Stuart. 4th Intern. Conf. Atom. Spectros., Toronto, November 1973.
32. Omang, S. V. *Anal. Chim. Acta, 53,* 415 (1971).
33. Chau, Y. K. and H. Saitoh. *Environ. Sci. Technol., 4,* 839 (1970).
34. Kalb, G. W. *Atomic Absorption Newsletter, 9,* 84 (1970).
35. Fishman, M. J. *Anal. Chem., 42,* 1462 (1970).
36. Brandenberger, H. and H. Bader. *Helv. Chim. Acta, 50,* 1409 (1967).
37. Brandenberger, H. and H. Bader. *Atomic Absorption Newsletter, 7,* 53 (1968).
38. Doherty, P. E. and R. S. Dorsett. *Anal. Chem., 43,* 1887 (1971).
39. Uthe, J. F., F. A. J. Armstrong, and M. P. Stainton. *J. Fisheries Res., 27,* 805 (1970).
40. Jeffus, M. T. and J. S. Elkins. *J. Assoc. Offic. Anal. Chemists, 53,* 1172 (1970).

41. Malaiyandi, M. and J. P. Barrette. *Anal. Lett.*, *3*, 579 (1970).

42. Munns, R. K. and D. C. Holland. *J. Assoc. Offic. Anal. Chemists*, *54*, 202 (1971).

43. Magos, L. *Analyst*, *96*, 847 (1971).

44. McCulloch, D. S., D. H. Peterson, T. J. Conomos, K. W. Leong, and P. R. Carlson. *Trans. Am. Geophys. Union*, *52*, 361 (1971).

45. Joensuu, O. I. *Appl. Spectrosc.*, *25*, 526 (1971).

46. Winefordner, J. D. and R. A. Staab. *Anal. Chem.*, *36*, 165 (1964).

47. Mansfield, J. M., J. D. Winefordner, and C. Veillon. *Anal. Chem.*, *8*, 1049 (1965).

48. Browner, R. F., R. M. Dagnall, and T. S. West. *Talanta*, *16*, 75 (1969).

49. Muscat, V. I. and T. J. Vickers. *Anal. Chim. Acta*, *57*, 23 (1971).

50. Muscat, V. I., T. J. Vickers, and A. Andren. *Anal. Chem.*, *44*, 218 (1972).

51. April, R. W. and D. N. Hume. *Science*, *170*, 849 (1970).

52. Bache, C. A. and D. J. Lisk. *Anal. Chem.*, *43*, 950 (1971).

53. Cooke, D. O., R. M. Dagnall, B. L. Sharp, and T. S. West. *Spect. Lett.*, *4*, 91 (1971).

54. Ling, C. *Anal. Chem.*, *40*, 1876 (1968).

55. Bailey, B. W. and F. C. Lo. *Anal. Chem.*, *43*, 1525 (1971).

56. Matusiak, W., M. Cefola, L. D. Cortivo, and C. J. Umberger. *Anal. Biochem.*, *8*, 463 (1964).

57. Weiss, H. V. and T. E. Crozier. *Anal. Chim. Acta*, *58*, 231 (1972).

58. Becknell, D. E., R. H. Marsh, and W. Allie, Jr. *Anal. Chem.*, *43*, 1230 (1971).

59. Smith. H. *Anal. Chem.*, *35*, 635 (1963).

60. Rottschafer, J. M., J. D. Jones, and H. B. Mark, Jr. *Environ. Sci. Technol.*, *5*, 336 (1971).

61. Pillay, K. K. S., C. C. Thomas, Jr., J. A. Sondel, and C. M. Hyche. *Anal. Chem.*, *43*, 1419 (1971).

62. Rook, H. L., T. E. Gills, and P. D. LaFleur. *Anal. Chem.*, *44*, 1114 (1972).

63. Westöö, G. *Acta Chem. Scand.*, *20*, 2131 (1966).

64. Westöö, G. *Acta Chem. Scand.*, *21*, 1790 (1967).

65. Kharkar, D. P., K. K. Turekian, and K. K. Bertine. *Geochim. Cosmochim. Acta*, *32*, 285 (1968).

66. Preston, A., D. F. Jefferies, J. W. R. Dutton, B. R. Harvey, and A. K. Steele. *Environ. Poll.*, *3*, 69 (1972).

67. Baseline Studies of Pollutants in the Marine Environment and Research Recommendations, The IDOE Baseline Conference, May 24-26, 1972, New York, 1972.

68. Slavin, W. *Atomic Absorption Spectroscopy*, John Wiley and Sons, New York, 1968.

69. Belcher, R., R. M. Dagnall, and T. S. West. *Talanta*, *11*, 1257 (1964).

70. West, F. K., P. W. West, and F. A. Iddings. *Anal. Chem.*, *78*, 1566 (1966).
71. West, F. K., P. W. West, and F. A. Iddings. *Anal. Chim. Acta*, *37*, 112 (1967).
72. Chao, T. T., E. A. Jenne, and L. M. Heppting. U.S. Geol. Survey Prof. Paper 600-D, U.S. Gov't Printing Office, Washington, D.C., 1968, pp. D13-D15.
73. Dyck, W. *Anal. Chem.*, *40*, 454 (1968).
74. West, F. K., P. W. West, and T. B. Ramakrishna. *Environ. Sci. Technol.*, *1*, 717 (1967).
75. Lee, R. F. and W. F. Pickering. *Talanta, 18*, 1083 (1971).
76. Chao, T. T., M. J. Fishman, and J. W. Ball. *Anal. Chim. Acta*, *47*, 189 (1969).
77. Handley, T. H. and J. A. Dean. *Anal. Chem.*, *32*, 1878 (1960).
78. West, T. S. and X. K. Williams. *Anal. Chem.*, *40*, 335 (1968).
79. Chao, T. T. and J. W. Ball. *Anal. Chim. Acta, 54*, 166 (1971).
80. Beamish, F. E., C. L. Lewis, and J. C. Van Loon. *Talanta, 16*, 1 (1969).
81. Lai, M. G. and H. V. Weiss. *Anal. Chem., 34*, 1012 (1962).
82. Johnson, D. L. *Environ. Sci. Technol.*, *5*, 411 (1971).
83. Wiersman, J. H. and G. F. Lee. *Environ. Sci. Technol.*, *5*, 1203 (1971).
84. Hashimoto, Y., J. Y. Hwang, and S. Yanagisawa. *Environ. Sci. Technol.*, *4*, 157 (1970).
85. Oelschlager, V. W. and K. H. Menke. *Ernahrungswiss, 9*, 216 (1969).
86. Angino, E. E., L. M. Magnuson, T. C. Waugh, O. K. Galle, and J. Bredfeldt. *Science, 168*, 389 (1970).
87. McGee, W. W. and J. D. Winefordner. *Anal. Chim. Acta, 37*, 429 (1967).
88. Ando, A., M. Suzuki, K. Fuwa, and B. L. Vallee. *Anal. Chem., 41*, 1974 (1969).
89. McClellan, B. W. and J. C. Chambers. 4th Intern. Conf. Atom. Spectros., Toronto, November 1973.
90. Severne, B. C. and R. R. Brooks. *Anal. Chim. Acta, 58*, 216 (1972).
91. Fernandez, F. J. and D. C. Manning. *Atomic Absorption Newsletter, 10*, 86 (1971).
92. Chu, R. C., G. P. Barron, and P. A. W. Baumgarner. *Anal. Chem., 44*, 1476 (1972).
93. Goulden, P. D. and P. Brooksbank. 4th Intern. Conf. Atom. Spectros., Toronto, November 1973.
94. Christian, G. D. 4th Intern. Conf. Atom. Spectros., Toronto, November 1973.
95. Manning, D. C. *Atomic Absorption Newsletter, 10*, 123 (1971).

96. Danchik, R. S. and D. F. Boltz. *Anal. Lett.*, *1*, 901 (1968).

97. Ihnat, M. 4th Inter. Conf. Atom. Spectros., Toronto, November 1973.

98. Dagnall, R. M., K. C. Thompson, and T. S. West. *Talanta*, *14*, 557 (1967).

99. Cresser, M. S. and T. S. West. *Spect. Lett.*, *2*, 9 (1969).

100. Lichte, F. E. and R. K. Skogerboe. *Anal. Chem.*, *44*, 1480 (1972).

101. Dean, J. A. and R. E. Fues. *Anal. Lett.*, *2*, 105 (1969).

102. Portman, J. E. and J. P. Riley. *Anal. Chim. Acta*, *31*, 509 (1964).

103. Ray, B. J. and D. L. Johnson. *Anal. Chim. Acta*, *63*, 108 (1970).

104. Blotcky, A. J., L. J. Arsenault and E. P. Rack. *Anal. Chem.*, *45*, 1056 (1973).

105. Myers, D. J. and J. Osteryoung. *Anal. Chem.*, *45*, 207 (1973).

106. Cervenka, A and M. Korbova. *Chem. Listy.*, *49*, 1158 (1955).

107. Raihle, J. A. *Environ. Sci. Technol.*, *6*, 621 (1972).

108. Young, J. W. and G. D. Christian. *Anal. Chim. Acta*, *65*, 127 (1973).

109. Friberg, L. T., M. Piscator, and G. F. Nordberg. *Cadmium in the Environment*, Chemical Rubber Co. Press, Cleveland, 1971.

110. Yamagata, N. and I. Shigematsu. Proc. Intern. Symp. on Hydrogeochemistry and Biogeochemistry, September 7-12, 1970, Tokyo.

111. Abdullah, M. I., L. G. Royle, and A. W. Morris. *Nature*, *235*, 158 (1972).

112. Smith, J., R. Nicholson, and J. P. Moore. *Nature*, *232*, 393 (1971).

113. Pulido, P., K. Fuwa, and B. L. Vallee. *Anal. Biochem.*, *14*, 393 (1966).

114. Dagnall, R. M., T. S. West, and P. Young. *Talanta*, *13*, 803 (1966).

115. Silvester, M. D. and W. J. McCarthy. *Anal. Lett.*, *2*, 305 (1969).

116. Silvester, M. D. and W. J. McCarthy. *Spectrochim. Acta*, *25B*, 229 (1970).

117. Livingston, H. D., H. Smith, and N. Stojanovic. *Talanta*, *14*, 505 (1967).

118. Brand, M. J. D., J. J. Militello, and G. A. Rechnitz. *Anal. Lett.*, *2*, 523 (1969).

119. Chow, T. J. *Nature*, *225*, 295 (1970).

120. Chow, T. J. and J. L. Earl. *Science*, *169*, 557 (1970).

121. Lazrus, A. L., E. Lorange, and J. P. Lodge, Jr. *Environ. Sci. Technol.*, *4*, 55 (1970).

122. Patterson, C. in *Impingement of Man on the Oceans* (D. W. Hood, ed.), John Wiley and Sons, New York, 1971, pp. 245-258.

123. Murozumi, N., T. J. Chow, and C. Patterson. *Geochim. Cosmochim. Acta, 33,* 1247 (1969).

124. Chow, T. J. and C. Patterson. *Earth Plan., 1,* 397 (1966).

125. Garrels, R. M. and C. L. Christ. *Solutions, Minerals and Equilibria,* Harper and Row, New York, 1965.

126. Gregor, C. D. *Environ. Sci. Technol., 6,* 278 (1972).

127. Willis, J. B. *Anal. Chem., 34,* 614 (1962).

128. Yeager, D. W., J. Cholak, and E. W. Henderson. *Environ. Sci. Technol., 5,* 1020 (1971).

129. Kahn, H. L. and J. S. Sebestyen. *Atomic Absorption Newsletter, 9,* 33 (1970).

130. Delves, H. T. *Analyst, 95,* 431 (1970).

131. Hwang, J.Y., P. A. Ullucci, S. B. Smith, Jr., and A. L. Malenfant. *Anal. Chem., 43,* 1319 (1971).

132. Loftin, H. P., C. M. Christian, and J. W. Robinson. *Spect. Lett., 3,* 161 (1970).

133. Woodriff, R. and J. F. Lech. *Anal. Chem., 44,* 1323 (1972).

134. Browner, R. F., R. M. Dagnall, and T. S. West. *Anal. Chim. Acta, 50,* 375 (1970).

135. Sychra, V. and J. Matousek. *Talanta, 17,* 363 (1970).

136. Prakash, N. J. and W. W. Harrison. *Anal. Chim. Acta, 53,* 421 (1971).

137. Donega, H. M. and T. E. Burgess. *Anal. Chem., 42,* 1521 (1970).

138. Rechnitz, G. A. and N. C. Kenny. *Anal. Lett., 3,* 259 (1970).

139. Whitnack, G. C. and R. Sasselli. *Anal. Chim. Acta, 47,* 267 (1969).

140. Hislop, J. S. and D. R. Williams. *Analyst, 96,* 78 (1972).

APPENDICES

APPENDIX 1

ABBREVIATIONS, UNITS AND SYMBOLS USED

(Standard abbreviations and common names
of chemicals given as Appendix 2.)

ABBREVIATIONS

AA	Atomic absorption
AE	Atomic emission
AF	Atomic fluorescence
AS	Atomic spectroscopy
Const	A constant
EDT	Electrodeless discharge tube
HC	Hollow cathode (lamp)
UV	Ultraviolet
aq	Aqueous phase
h.f.	High frequency
org	Organic phase
ppb	Parts per (U.S.) billion (10^9)
ppm	Parts per million (10^6)
rms	Root-mean-square
r.f.	Radio frequency

UNITS

Å	Ångstrøm unit (10^{-8} cm; 10^{-1} nm)
ℓ	Liter
m	Meter
μg	Microgram (10^{-6} g)
ng	Nanogram (10^{-9} g)
pg	Picogram (10^{-12} g)

SYMBOLS

A	Absorbance; or anion
B	Einstein probability coefficient (absorption)

311

C	Concentration; or coefficient of variation
D	Extraction coefficient
E	Extraction efficiency
E_D	Dissociation energy (ev)
$E_{D.Ox.}$	Dissociation energy of oxide (RO; ev)
E_i	Ionization energy (ev)
E_j	Excitation energy of state j
E_t	rms thermal noise voltage
Eh	Redox potential
F	Flow rate as defined; or faraday
ΔF	Detector circuit frequency band width
H	Hydrogen
HL	Chelating agent
ΔH	Heat of reaction (enthalpy change; kcal/mole)
I	Ionic strength
I_A	Absorbed radiation intensity (integrated; energy per unit time per unit energy)
I_E	Emission intensity
I_F	Fluorescence intensity
I_O	Integrated intensity of (narrow) line source
$I_{O\lambda}$	Intensity of continuum source at λ
I_S	rms shot-noise current
I_{tot}	Total circuit current
I_λ	Intensity of source emission at λ
I_λ^B	Emission intensity of a black body at λ
K	Temperature (°K)
$\overline{K}$	Average atomic absorption coefficient over source line width
K_D	Distribution coefficient (resin ion-exchange)
K_d	Dissociation constant
K_i	Ionization constant
K_S	Selectivity coefficient (resin ion-exchange)
K_λ	Atomic absorption coefficient at λ
K_ν	Atomic absorption coefficient at ν
L^-	Organic ligand
M	Molarity (moles solute/liter)
N	Avogadro number
N_{equil}	Equilibrium value of N_t
N_j	Number of atoms in excited state j
N_O	Number of ground-state atoms
N_t	Number of atoms in cell at time t
N_{tot}	Total number of atoms in discrete sample
O	Oxygen
P	Thermodynamic partition coefficient
P_A	Total power absorbed ($I_A \cdot$ cell area)

P_F	Total fluorescence power emitted ($I_F \cdot$ cell area)
P_{RA}	Vapor pressure of compound RA
Q_N	Integrated signal-pulse area
Q_{RA}	Rate of evaporation of compound RA
R	Gas constant; or unspecified cation; or resistance
$(R)_a$	Thermodynamic activity of metal ion R^{n+} in phase a
$[R]_a$	Concentration of metal in phase a
$R°$	Ground-state atom
$R°*$	Excited atom
R^{n+}	Metal ion
Re	Reynolds number
S	Variance (σ^2)
T	Temperature (°C); or residence time
T_{rel}	Relative residence time
V	Velocity; or volume as defined
X	Chemical species (generally in flame)
Y	Fraction of fluorescence radiation measured (ster)
Z	Atomic partition function
a	Unspecified phase; or activity
b	Cell path-length
c	Velocity of light
e	Electron charge
f	Oscillator strength; or function
g_j	Statistical weight of excited state
h	Planck constant
k	Boltzmann constant
l_c	Length of capillary
$\ln$	Log_e
log	Log_{10}
m	Milli (10^{-3})
m_a	Atomic weight
m_{RA}	Molecular weight of compound RA
m_e	Electron mass
n	Nano (10^{-9}); or unspecified integer
p	Pico (10^{-12})
pH	$-\text{Log } a_{H^+}$
pϵ	$-\text{Log } a_\epsilon$
r	Radius as defined
r_R	Reaction rate
s	Spectral band width of monochromator
t	Time; or student "t"
x	Any measured parameter
x_o	True value of measured parameter

$\bar{x}$	Arithmetic mean of n measurements of x
Ω	Solid angle of fluorescence radiation measured (ster)
α	Fraction of absorbed radiation (I_A/I_O)
α_d	Degree of dissociation
α_i	Degree of ionization
β	Degree of atomization
γ	Sensitivity (conc^{-1})
ε	Electron
η	Viscosity coefficient
λ	Wavelength (generally in Å)
$\Delta\lambda_D$	Doppler broadening half-width
$\Delta\lambda_a$	Absorption line half-width
$\Delta\lambda_b$	Source line half-width
$\Delta\lambda_c$	Collisional broadening half-width
μ	Micro (10^{-6}); or micron (10^{-6} m)
μ_a	Chemical potential (in phase a)
$\mu°$	Standard state chemical potential
ν	Frequency
ρ	Density as defined
σ	Standard deviation
σ_b	Standard deviation of blank or background
τ	Lifetime of excited atom ($\sim 10^{-8}$ sec)
τ_c	Residence time of atoms within cell
τ_{cir}	Time constant of recording circuit
τ_s	Signal recording time
τ_T	Transfer time of atoms to cell
ϕ	Quantum efficiency
ψ	Flame gas expansion factor

APPENDIX 2

COMMON NAMES AND ABBREVIATIONS OF CHEMICALS

APDC	Ammonium pyrrolidene dithiocarbamate (carbodithioate)
Bathocuproine	2,9,-dimethyl-4,7-dephenyl-1,10-phenanthroline
Bathophenanthroline	4,7-diphenyl-1,10-phenanthroline
Cupferron	Ammonium salt of n-nitrosophenyl-hydroxylamine
Cuprethol	Bis (2-hydroxyethyl) dithiocarbamate
Cuproine	2,2'-biquinoline
DEDTC	(Sodium) diethyldithiocarbamate; cupral; DDTC
DADDC	Diethylammonium diethyldithiocarbamate; DADDTC
Diaminobenzidine	Diamino-3,3'-benzidine
Dithiol	Toluene-3,4-dithiol
Dithizone	Diphenylthiocarbazone
EDTA	Ethylenediamine tetraacetic acid
HAA	Acetylacetone
Heptoxime	1,2-cycloheptanedionedioxime
IOTG	Isooctylthioglycolate
MIBK	Methyl isobutyl ketone
Neocuproine	2,9-dimethyl-1,10-phenanthroline
Nitroso-R	1-nitroso-2-hydroxy-3,6-naphthalene disodium sulfonate
NTA	Nitrilotriacetic acid
Oxine	8-hydroxyquinoline; quinolinol
PAQA	Pyridine-2-aldehyde-2-quinolyaldehyde
Phenylfluorone	2,3,7-trihydroxy-9-phenyl-6-fluorone
PHTTT	3-propyl-5-hydroxy-5-D-arabinotetra-hydroxybutyl-3-thiazolidine-2-thione
QXDT	Quinoxaline-2,3-dithiol
Selenazone	Selenium analog of dithizone
S-oxime	Salicylaldoxime
TEPA	Tetraethylenepentamine
TFA	Trifluoroacetylacetone

Thorin   (Sodium) hydroxy-2-disulfo-3,6-
                naphthylazo-1,2-benzenearsonic acid
Tiron   (Sodium) 1,2-dihydroxybenzene-3,5-
                disulfonate
TOTP   Triisoclylthiophosphate
Tripyridine   2,2',2"-tripyridine (terpyridine)
Zincon   2-carboxy-2'-hydroxy-5'-sulfoformazl-
                benzene

APPENDIX 3

STANDARD STOCK SOLUTIONS
(1000 MG/L)

Ag   Dissolve 1.575 g $AgNO_3$ in water and dilute to 1 ℓ. Store in amber glass.

As   Dissolve 1.320 g As_2O_3 in 3 ml 8 M HCl and dilute to 1 ℓ.

Au   Dissolve 1.000 g Au in 10 ml hot HNO_3 by dropwise addition of HCl; boil; dilute to 1 ℓ. Store in amber glass.

Bi   Dissolve 1.000 g Bi in 8 ml 10 M HNO_3; boil gently to expel brown fumes; dilute to 1 ℓ.

Cd   Dissolve 1.000 g Cd in 10 ml of 2 M HCl and dilute to 1 ℓ.

Co   Dissolve 1.000 g Co in 10 ml of 2 M HCl and dilute to 1 ℓ.

Cr   Dissolve 2.829 g $K_2Cr_2O_7$ in water and dilute to 1 ℓ.

Cu   Dissolve 3.929 g $CuSO_4 5H_2O$ in water and dilute to 1 ℓ.

Fe   Dissolve 1.000 g Fe wire in 20 ml of 5 M HCl and dilute to 1 ℓ.

Ga   Dissolve 1.000 g Ga in 50 ml of 2 M HCl and dilute to 1 ℓ.

Ge   Dissolve 1.441 g GeO_2 and 50 g oxalic acid in 100 ml of water and dilute to 1 ℓ.

Hf   Add 10 ml of 9 M H_2SO_4 to 1.000 g Hf in platinum dish and HF dropwise to complete dissolution. Dilute to 1 ℓ with 10% H_2SO_4.

Hg   Dissolve 1.000 g Hg in 10 ml of 5 M HNO_3 and dilute to 1 ℓ.

317

In   Dissolve 1.000 g In in 50 ml of 2 M HCl and dilute to 1 ℓ.

Ir   Dissolve 2.465 g Na_3IrCl_6 in water and dilute to 1 ℓ.

Mn   Dissolve 3.070 g (dry) $MnSO_4 \cdot H_2O$ in water and dilute to 1 ℓ.

Mo   Dissolve 2.043 g $(NH_4)_2MoO_4$ in water and dilute to 1 ℓ.

Nb   Add 20 ml of HF to 1.000 g Nb in platinum dish and heat to complete dissolution. Cool; add 40 ml H_2SO_4; evaporate to fumes; cool; dilute to 1 ℓ with 8 M H_2SO_4.

Ni   Dissolve 1.000 g Ni in 10 ml hot HNO_3, cool and dilute to 1 ℓ.

Os   Dissolve 1.336 g OsO_4 water and dilute to 1 ℓ.

Pb   Dissolve 1.599 g $Pb(NO_3)_2$ in HNO_3 and dilute to 1 ℓ.

Pd   Add 1.000 g Pd to 10 ml HNO_3 and dissolve by dropwise addition of HCl to hot solution; dilute to 1 ℓ.

Pt   Dissolve 1.000 g Pt in 40 ml hot aqua regia, evaporate close to dryness, add 10 ml HCl and repeat evaporation step; add 10 ml HCl and dilute to 1 ℓ.

Re   Dissolve 1.000 g Re in 10 ml 8 M HNO_3 in an ice bath. Dilute to 1 ℓ.

Rh   Add 1.000 g Rh to a glass tube, add 20 ml HCl and 1 ml $HClO_4$. Seal tube and place in oven at 300° C for 24 hr. Cool, break open tube, transfer solution to a volumetric flask and dilute to 1 ℓ.

Ru   Dissolve 1.317 g RuO_2 in 15 ml HCl and dilute to 1 ℓ.

Sb   Dissolve 1.000 g Sb in 10 ml HNO_3 plus 5 ml HCl and dilute to 1 ℓ.

Se   Dissolve 1.4050 g SeO_2 in water and dilute to 1 ℓ.

Sn   Dissolve 1.000 g Sn in 15 ml warm HCl and dilute to 1 ℓ.

Ta   Add 20 ml HF to 1.000 g Ta in platinum dish, heat gently to complete dissolution. Cool, add 40 ml H_2SO_4 and evaporate to fumes. Cool and dilute to 1 ℓ with H_2SO_4.

Te   Dissolve 1.2508 g TeO_2 in 10 ml HCl and dilute to 1 ℓ.

Ti   Dissolve 1.000 g Ti in 10 ml of H_2SO_4 with dropwise addition of HNO_3 and dilute to 1 ℓ with dilute H_2SO_4.

Tl   Dissolve 1.303 g $TlNO_3$ in water and dilute to 1 ℓ.

V   Dissolve 2.296 g NH_4VO_3 in 100 ml water with 10 ml HNO_3 and dilute to 1 ℓ.

W   Dissolve 1.794 g $Na_2WO_4 \cdot 2H_2O$ in water and dilute to 1 ℓ.

Zn   Dissolve 1.000 g Zn in 10 ml HCl and dilute to 1 ℓ.

Zr   Dissolve 3.533 g $ZrOCl_2 \cdot 8H_2O$ in 50 ml 2 M HCl and dilute to 1 ℓ.

APPENDIX 4

HEAVY METALS ELEMENTAL DATA

Symbol	Element	Atomic Number	Atomic Weight
Ag	Silver	47	107.88
As	Arsenic	33	74.91
Au	Gold	79	197.00
Bi	Bismuth	83	209.00
Cd	Cadmium	48	112.41
Co	Cobalt	27	58.94
Cr	Chromium	24	52.01
Cu	Copper	29	63.54
Fe	Iron	26	55.85
Ga	Gallium	31	69.72
Ge	Germanium	32	72.60
Hf	Hafnium	72	178.60
Hg	Mercury	80	200.61
In	Indium	49	114.76
Ir	Iridium	77	192.20
Mn	Manganese	25	54.94
Mo	Molybdenum	42	95.95
Nb	Niobium	41	92.91
Ni	Nickel	28	58.69
Os	Osmium	76	190.20
Pb	Lead	82	207.21
Pd	Palladium	46	106.70
Pt	Platinum	78	195.23
Re	Rhenium	75	186.31
Rh	Rhodium	45	102.91
Ru	Ruthenium	44	101.10
Sb	Antimony	51	121.76
Se	Selenium	34	78.96
Sn	Tin	50	118.70
Ta	Tantalum	73	180.95
Te	Tellurium	52	127.61
Ti	Titanium	22	47.90
Tl	Thallium	81	204.39
V	Vanadium	23	50.95
W	Tungsten	74	183.92
Zn	Zinc	30	65.38
Zr	Zirconium	40	91.22

INDEX

INDEX

325